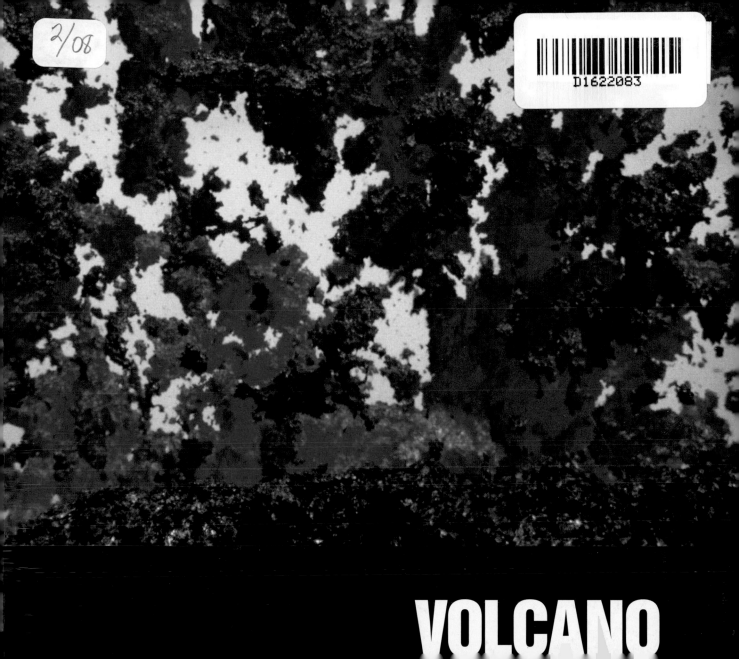

VOLCANO

VOLCANO
A VISUAL GUIDE

DONNA O'MEARA
FOREWORD BY
PROFESSOR AUBREY MANNING

FIREFLY BOOKS

CONTENTS

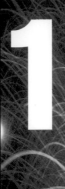

THE BIG BANG

1

2

CONES CRATERS & CALDERAS

FLOWS FOUNTAINS & LAKES

3

Snow Tower, Sauna Cave, Mt. Erebus, Antarctica

THE SNOW TOWER marks the location of an underground ice cave named Sauna Cave. Heat from a large volcanic steam vent sitting under the cave melted the hollow into the ice. As the ice melts, water dribbles down through cracks onto hot rocks and is recycled into steam keeping the inside of the cave warm. Visitors to the caves use them as soothing saunas. The tower itself formed from water vapour that froze around an opening in the ceiling of the cave.

When I was but thirteen or so
I went into a golden land;
Chimborazo, Cotopaxi
Took me by the hand.

W. J. Turner's mysterious poem, "Romance", has always evoked a most powerful response from me. It is those strange, beautiful names and, above all, the images that they conjure: the perpetual snows and the craters whose heat and fumes reveal their link to the molten centre

FOREWORD

of the planet. Volcanoes are special because even as we take delight in their strange beauty, there must always be an undercurrent of fear. They force us to recognize that the gentle surface of the Earth on which we live is just a thin crust – the fires are below. Nevertheless, despite their obvious dangers, volcanic regions are often densely populated: many millions of people continue to live around active volcanoes. As they spew gas and fire and ash, volcanoes also create new soil, often rich in minerals brought up from below, which can mean great fertility for crops grown nearby. People are drawn to the slopes again and again, taking their chances and sometimes having to retreat and lose everything.

One of the great achievements of modern science has been the final understanding of how our planet 'works' that happened barely 50 years ago, out of which came a key recognition of the part played by volcanoes. When I was a student, plate tectonics ('continental drift' as it was known to us) was still regarded by many scientists as a crazy idea. But perhaps the last crucial piece of evidence came on Good Friday 1964, with the great Alaskan earthquake. Geologists worked out that it was caused by part of the Pacific plate of the Earth's crust sliding beneath the North American plate. This subduction process is part of the cycle of the Earth's dynamic life. The plates can also slide past one another as in California or they can push against one another as occurs in Alaska, where the edge of one plate sinks below the other. New crust that forms at the mid-ocean ridges where molten magma bursts to the surface, can also move the plates apart. Water is carried down with the rocks, which melt, and some of this molten magma breaks through to the surfac, with dramatic results. Thus, volcanoes are an outlet, continually recycling the elements of the crust and the mantle, bringing up huge quantities of minerals and gasse. These gasses form and become active, sometimes briefly, sometimes for many millennia and their recharging of our atmosphere with carbon dioxide has through billions of years been one of the crucial mechanisms keeping Earth's temperatures moderate, with liquid water at the surface, thus giving our planet the ability to support life.

All of this wonderful story is described here in detail by Donna O'Meara in words and in photographs – the latter also being awe-inspiring. We are becoming so used to the amazing standards of modern film and photography that it takes a lot to surprise us now; but this book will do so nevertheless. It is salutary for all of us to contemplate the gigantic forces of the Earth where human endeavours are trivial. This book gives us the opportunity to do just that.

PROFESSOR AUBREY MANNING

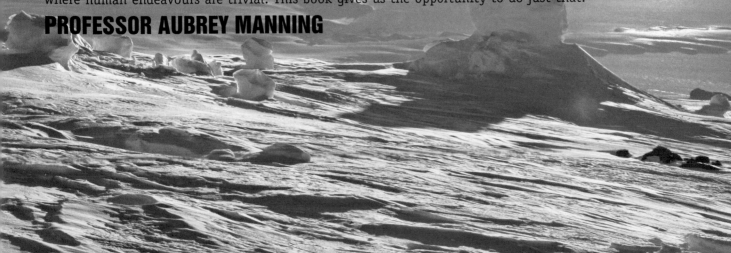

INTRODUCTION

Volcanoes are a smorsgasborg for the senses. Steam hisses, lava gurgles, flows gush and tinkle like broken glass, and blasts rumble and boom. Sulfury gasses tickle the nose with a pungent, but not completely offensive, odour. Whether majestically slumbering or thundering in eruption volcanoes are a visual treat.

Volcanoes are indelibly etched on the human psyche through myth, genetic memory and historical records. The earliest human recording of a volcano erupting is a colorful wall painting in central Turkey that dates to 6000BC. Greek mythology cites Italy's Mt. Etna as the home of Hephaestus, the god of fire and the Romans believed their own god of fire, Vulcan, resided in Vulcano in Italy's, Aeolian (Lipari) Islands. And in Nicaragua, only the loveliest of virgins were sacrificed into the boiling lava lake of Masaya Volcano to appease Chaciutique, its goddess of fire. Even the mythical city of Atlantis may have been submerged and destroyed by a volcano when a cataclysmic eruption of Santorin (Thera) volcano in Greece caused a tsunami (some believe Santorini itself is Atlantis).

Earth is peppered with more than 1,500 active volcanoes, two-thirds of which are in the Northern hemisphere. Of those 1,500 about 500 have erupted in modern times. Of course, volcanoes are not haphazardly scattered across Earth's surface. 80 per cent grow along the edges of vast tectonic plates riding the Earth's partially molten mantle, pulling away from each other or crashing into each other, forcing magma up and out along their boundaries. Where tectonic plates diverge and pull apart, ridges of volcanoes form, such as occurs at the Mid-Atlantic Ridge. Where Earth's plates push together, converge and are forced under each other (subduction), long swathes of volcanoes are found, such as the colossal Pacific Ring of Fire. This necklace of over 1,000 volcanoes comprises nearly two thirds of the known terrestrial volcanoes on Earth and stretches up the northwest coast of the United States, through Alaska's Aleutian Islands, and down into Japan. Convergent plates also build volcanic islands and island arcs, like those in Indonesia and when Earth's crust stretches and thins, intra-plate volcanism can occur, such as in the African Rift Valley and at various volcanic hot spots. These occur where plumes of stationary magma, smoulder under sliding plates, creating 'conveyor belt' volcanoes like those in the Cape Verde Islands, far from the edges of tectonic plates.

The study of volcanoes is a new and emerging science and scientists are not yet in agreement about classifying volcanoes or about their lifespans. Generally the definition of an active volcano is one that has erupted within the last 10,000 years; however, super-volcanoes like Yellowstone in the USA, have eruptive lifespans that last for hundreds of thousands, and possibly, millions of years, so we cannot be sure they will not erupt again.

Earth is not the only planet in our solar system to have volcanoes and several other solar-system bodies either are volcanic or show signs of volcanic activity. As astronomers are just beginning to glimpse other planets around stars other than our Sun, it may be a matter of time before we have the technology to discover new worlds in new solar systems with volcanoes just like those on Earth.

DONNA O'MEARA

WHEN VOLCANIC PRODUCTS EMERGE AT THE SURFACE, it is defined as a volcanic eruption. Erupting volcanoes are many different sizes and shapes, which in turn are vastly affected by the type of volcanic products they emit – ash, gas or lava, or a combination of either of these. Nearly 8,000 independent eruptions from Earth's volcanoes have been witnessed and scientifically recorded but many others in inaccessible regions or along the ocean floor go undocumented. Much eruptive activity is named after well-known volcanoes which are known for a certain type of 'behaviour': these eruptions are termed Strombolian, Vulcanian, Vesuvian and Pelean- and Hawaiian-type eruptions. Volcanoes can exhibit one distinct type of eruption or many types combined, and when volcanoes erupt, we have a magical view inside the Earth's molten interior.

Eruptions are either effusive (fluid) lavas like the ongoing lava flows at Hawaiian volcanoes, or dangerous and explosive like the blasts of molten rock, gas, ash and other pyroclasts that occurred at Sumatra's Tambora volcano in 1815. Eruptions such as this are deadly, and this one was in fact responsible for as many as 90,000 deaths. Some eruptions occur for just a few minutes and others spew their products for decades. Within these two main categories of effusive and explosive eruptions, there are many sub-designations: eruptions where gasses mix with gritty pulverized rock forming tall dark ash clouds seen for many kilometers such as were witnessed at Mt. St. Helens, USA, in 1980, flank fissure eruptions with lava oozing from long horizontal cracks

on the side of a volcano such as at Piton de la Fournaise volcano on Réunion island. Then there are the ground hugging lethally hot avalanches of volcanic debris called pyroclastic flows like those in 1991 at Pinatubo volcano in the Philippines. When rising magma encounters ground water, enormous phreatic (steam) eruptions can happen, like those at Poas volcano in Costa Rica. Eruptions can release suffocating gasses into the atmosphere, cause tsunamis and floods, trigger earthquakes and unleash ravaging mudflows and rock slides.

Despite all of the above, volcanic eruptions aren't entirely bad. If not for volcanoes humans might not exist. The air that we breath contains erupted volcanic vapours and the oceans in which life began were formed from condensed volcanic steam. Eruptions add rich soil to the planet's surface, create new land masses and extend shorelines into the sea. Eruptions provide geothermal resources, therapeutic hot springs and renewable energy. Not to mention their mysterious inspirational beauty. One day we may learn enough about volcanoes to be able to forecast when a specific eruption might occur. Then, humanity will be one step closer to living in harmony with these magnificent mountains of fire.

THE BIG BANG

1

Icelandic fissure eruption, Eldfell, Iceland

ICELANDIC ERUPTIONS SPEW exceptionally fluid basalt lavas from long, narrow cracks, or fissures, in the Earth. These flood-like lava flows create vast lava plateaus. On January 21, 1973, the fishing village of Vestmannaeyjar, Heimaey island, Iceland, perched on the side of Helgafell volcano, was rattled by dozens of earthquakes forcing residents to be evacuated by air. Helgafell hadn't erupted for 5,000 years and was considered extinct. 48 hours later 100m- / 328ft-high lava fountains ripped open a 1.6km- / 1 mile-long fissure, gushing lava flows raced towards the sea, burning homes in their path and a massive ash cloud 60km / 37 miles long and 12km / 7.5 miles wide blasted 7,000m / 22,965ft into the sky. When the eruption abated a 222m- / 728ft-high volcanic cone, named Eldfell, stood where a flat meadow had been.

Flank vent eruption, Piton de la Fournaise, Réunion

FLANK VENT ERUPTIONS are caused when magma moving under a volcano causes shallow earthquakes that crack, weaken and splinter the flanks (sides) down toward the volcano's magma chamber, forming what are called conduit vents that allow lava and gasses to escape. If lava piles up around a flank vent, small cones, called parasitic cones are created, and Piton de la Fournaise is dotted with them. The first documented report of activity at Piton de la Fournaise, (meaning Peak of the Furnace), in the western Indian Ocean, was in 1644. Today Piton leads the pack with the world's most recorded flank vent eruptions, totalling more than forty-five. This shield volcano is monitored very closely by the Institut de Physique du Globe de Paris, who run a nearby observatory.

Hawaiian-type eruption, Kilauea, Hawaii, USA

14   VOLCANOES ARE CLASSIFIED into several categories, with the Hawaiian kind being named after the typical Hawaiian
eruption, which consists of fluid outpourings of runny, basalt lava rather than big explosions of debris, rocks and
ash. Lava from Hawaiian eruptions can spew from linear or central vents, creating streams and rivers of glowing
molten rock that flow down the slope, pool in old craters and cover acres of land beyond – the low gas and silica
content of the lava keeps it flowing smoothly. Kilauea, a shield volcano on the island of Hawaii, 'The Big Island', is
the most active volcano on Earth and Hawaii's youngest volcano. Its eruptions have smothered and covered 75sq
km / 29sq miles, incinerated over 180 homes and, in 1990, buried the historic coastal town of Kalapana.

Strombolian eruption, Stromboli, Italy

STROMBOLIAN ERUPTIONS are named after Stromboli volcano in the Aeolian island chain, north-east of Sicily. Strombolian eruptions are noted for launching luminous lava, cinders, ash, lava 'bombs' and blocks, collectively called pyroclasts, skyward for hundreds of meters. Since Strombolian magma is generally stickier and more viscous than Hawaiian magma, gas bubbles can't easily escape and they explode as they rise in the vent, hurling out clots of molten lava in glorious, incandescent arcs. Stromboli volcano, about 4.8km / 3 miles in diameter, and jutting out into the Tyrrhenian Sea, has been continuously active for over 2,000 years and visitors flock to see 'The Lighthouse of the Mediterranean'. The largest recent eruption here occurred in 1930, killing three island residents.

Vulcanian eruption, Sakura-jima, Japan

A VULCANIAN ERUPTION (named after the Vulcano volcano in the Aeolian islands at the foot of Italy and Vulcan, the god of fire) is a medium-sized explosive blast of ash, gas and viscous lava blocks. When a volcano's magma contains a lot of gasses, it explodes. The force of the explosion creates a tall, debris column that rises straight up in the air, high above the volcano and then falls, and the accumulation is called tephra. One of Japan's most active strato-volcanoes, Sakura-jima, in Kagoshima Bay on the Japanese island of Kyushu, has erupted over one hundred times per year since the mid-1950s, frequently belching great dark clouds of smoke and ash from its summit crater. In 1995 an explosive eruption propelled an ash column 2,500m / 8,202ft into the atmosphere.

Ultra-Vulcanian eruption, Merapi, Java, Indonesia

WHEN A DANGEROUS STRATO-VOLCANO, such as Merapi (meaning mountain of fire) in Java, Indonesia, has a gigantic Vulcanian eruption, it is often termed ultra-Vulcanian. Mt. Merapi is extremely menacing. The problem being that the island of Java has one of the densest populations in the world, and the volcano is just 30km / 18.5 miles north of Yogyakarta city and its three million inhabitants. Mt. Merapi has generated billowing ash plumes over 10km / 6.2 miles high, causing mass evacuations. Eleven of Merapi's past eruptions, some of which have spread up to 5km / 3.1 miles away from the summit, have started fires and killed residents. Merapi is the most active volcano in Indonesia, and it is believed to be capable of even larger and more destructive eruptions.

Peléan eruption, Mayon, Philippines △

THE STEEP SYMMETRICAL CONE of Mayon volcano rises 2,462m / 8,077ft above Albay Gulf. It is the region's most active volcano with dense, thick lava, chiefly composed of rhyolite or andesite, creating Peléan eruptions (named after the lethal 1902 eruption of Mount Pclóc in the West Indies). These deadly eruptions begin when thick lava plugs up the volcano's vent and forms a dangerous volcanic dome. The pent up gasses finally cause the dome to collapse unleashing the deadly Peléan eruption. This flattens entire towns with its characteristic feature, a glowing, fiery avalanche of volcanic ash, gas and volcanic debris, called a pyroclastic flow. In 1968 a massive Peléan eruption of gas, dust, ash, and red-hot debris blew out of Mayon's central crater, then rushed down its sharp slopes at up to 160km / 99 miles per hour.

Plinian eruption, Pinatubo, Philippines ▷

PINATUBO IS A STRATO-VOLCANO dominating the island of Luzon in the northern Philippines. In 1991, after decades of slumber, it burst into life with one of the largest eruptions in the twentieth century. An enormous Plinian eruption (named after the Roman naturalist Pliny the Elder who described, and perished in, the 79AD eruption of Mt. Vesuvius in Italy) produced pyroclastic and smothering mudflows which covered towns and villages. Plinian eruptions begin with raging, uninterrupted gas eruptions, which can last months, and a tall, billowing eruption cloud containing huge volumes of volcanic pumice and ash. The cloud blasts tens of kilometers into the atmosphere, where fine ash and tephra is carried by the wind and deposited hundreds of kilometers away. Heavier volcanic rocks and heated debris plummet to Earth causing searing gas and ash to rush down the slopes and across the landscape.

Ash eruption, Popocatépetl, Mexico

ASH IS DEFINED AS TINY PIECES, measuring less than 2.8cm / 1.1in of pulverized rock blown from an explosive vent. Though it can be as fine as baby powder, it is invariably far more gritty and abrasive. Explosive eruptions, such as this one at Popocatépetl looming 5,426m / 17,801ft high above Mexico City, hurtle the fine particles to massive heights, and when they fall back to Earth and spread over the land in layers, geologists term it an 'air-fall deposit'. Large volcanic ash eruptions can drift up to 5,000km / 3,107 miles away, and have even clogged the engines of aeroplanes. Ash eruptions have occurred at Popocatépetl since at least Aztec times – its name is, in fact, an Aztec word, meaning smoking mountain.

Phreatic eruption, Laguna Caliente, Poas, Costa Rica △

POAS VOLCANO is one of the most active in Costa Rica. It sits just outside the capital city of San Juan and is known for its phreatic eruptions (or phreatic explosions) at Botos and Laguna Caliente, its twin crater lakes. Phreatic eruptions are explosions of steam, water and mud occurring when cool ground or surface water meets hotter volcanic rocks, magma, ash or debris (no new magma or lava is erupted). Laguna Caliente, the northern lake at Poas, is geothermally heated, extremely acidic and has seen numerous phreatic eruptions since the early 1800s. When Laguna Caliente at Poas erupts, it often squirts geyser-like columns of lake water skyward. Just a short drive from San José, the capital city, Poas's summit and slopes are covered with a wonderful cloud rainforest.

20

Crater lake eruption, Ruapehu, New Zealand ▽

NEW ZEALAND'S MT. RUAPEHU has a volcanic summit crater lake with the highest number of recorded crater-lake eruptions, currently forty-eight. When water in a crater lake gets too close to the hot magma underneath it, it can produce an explosive eruption. Ruapehu's eruptions also generate dangerous mudflows, called lahars, which cause havoc in, and flood, the surrounding valleys. On Christmas Eve, 1953, one such crater-lake eruption washed away a railway bridge and the Wellington-Auckland express train, killing 151 people. Frequent eruptions at Ruapehu have worn down its slopes. Today the volcano is equally well known for its popular ski resort.

Sub-glacial eruption, Grimsvötn, Iceland

GRIMSVÖTN VOLCANO IN ICELAND has produced more documented sub-glacial eruptions than any other volcano on Earth. These unique eruptions of lava under the ice, glaciers and ice caps, can be dramatic and produce 'puffy pillow' lava flows, usually only seen underwater. Heavy ice above such a lava flow can also flatten the top of it, leaving behind a smooth-topped mountain. About one-third of all the volcanoes capable of sub-glacial eruptions are in Iceland. Grimsvötn is a very active volcano buried under the 200m / 656ft thick Vatnajokull ice shelf. When it erupts, hot lava quickly melts the ice, releasing huge amounts of water which produce racing tides with flow rates faster than the River Amazon's. Grimsvötn's first recorded eruption was in 1332, and it has erupted about fifty times since.

Central crater eruption, Aso, Japan

SOMETIMES GEOLOGISTS DESCRIBE volcanic eruptions according to where on the volcano the eruption occurred and more than half of all volcanoes on Earth have central crater eruptions. A vent is a direct line from the opening at the Earth's surface to the magma storage area, or magma chamber and a central crater eruption occurs when volcanic materials erupt through the central vent, which is located at a volcano's summit caldera. Mt. Aso, in Japan, has had the most recorded central crater eruptions – over 151 – of any volcano. When a volcano has multiple eruptions, pyroclastic material and lava flows build up around the central vent and form the familiar cone or shield-shaped facade of a volcano. Aso's 24km- / 14.9 mile-wide caldera was formed about 300,000 years ago.

World's biggest volcano, Mauna Loa, Hawaii, USA

MAUNA LOA VOLCANO, one of five shield volcanoes on the island of Hawaii, is the largest mountain and the biggest volcano on Earth. Measured from its underwater base, in the Hawaiian trough, it is 160km / 100 miles long and rises about 10,700m / 35,000ft high, making it taller than Mt. Everest. Mauna Loa means long mountain in Hawaiian, and its gently sloping shield shape is about 97km / 60.2 miles wide. In total there's a massive 40,000cu km / 9,596.5cu miles of mountain. The summit caldera is named Mokuaweoweo, and there are two volcanic rift zones running northeast and southwest. Mauna Loa has erupted 15 times since 1900, the last time being in 1984 and many geologists feel it is overdue another eruption.

Pyroclastic eruption, Pinatubo, Philippines

WHEN MT. PINATUBO unleashed a pyroclastic flow in 1991, leaving deposits up to 220m / 722ft thick, the world was stunned. Before then, Pinatubo had been relatively unknown. Pyroclastic flows involve huge avalanches of scalding volcanic gas, ash and rocks racing down the slope at hurricane speeds of 200m / 656ft per second and temperatures of up to 1,000°C / 1,830°F. They cannot be outrun, and they flatten anyone or anything in their paths. Cars are melted, forests incinerated, cities buried and human beings carbonized in seconds. Pyroclastic flows are formed in several ways: when a massive eruption column collapses; when gravity makes a weakened part of a volcano collapse; or when thick, viscous lava clogs a volcanic vent causing a large, dense lava dome to inflate, and finally collapse or explode.

Pyroclastic eruption delta, Soufrière Hills, Montserrat

ANDESITIC LAVA IS THICK and clogs a volcano's vent, creating a lava dome. In 1997, when Castle Peak lava dome – nestling in the horseshoe-shaped 1km- / 0.6 mile-wide summit crater of Soufrière Hills (called English's Crater) – suddenly collapsed, it created a pyroclastic flow of debris which coursed down the White River channel and the resulting debris delta fanned out into the sea, on the southern side of the tranquil Caribbean island. Before 1995 there had been no recorded eruptions at Soufrière Hills, so it was extremely fortunate that the nearby population of nearly 5,000 people had been fore-warned by smaller gas and ash emissions, and had been evacuated to safety just three weeks prior.

Debris avalanche eruption, Augustine, Alaska, USA

AUGUSTINE VOLCANO, an island volcano in Kamishak Bay, Alaska, is the most active volcano in the eastern Aleutian arc. It is famed for its immense avalanches packed with debris, its eruptions and rock falls. A volcanic-debris avalanche occurs when the weight of the existing volcanic summit builds up, or a new eruption occurs, triggering a major collapse along the volcano's perimeter flanks, flinging blocks, rocks and ash down the slope. About a dozen debris avalanches have, over time, engulfed Augustine's base with a circular skirt of volcanic debris reaching into the surrounding ocean. Some of these rocks are over 40,000 years old. During the last two big flank collapses and debris-avalanche eruptions at Augustine the seismic force was equal to a magnitude 3.2 earthquake.

THIS IMAGE OF MERAPI VOLCANO, in Indonesia, shows the dramatic decoupling of the top and bottom layers of a pyroclastic flow. This moment of separation happens when heavier glowing rocks and hot chunks in the bottom section of the flow advance forward like a giant tsunami of fire, rapidly rolling along at ground level, while lighter ash and incinerating gaseous particles in the top layer float up in a furious, red-hot updraft, called a pyroclastic surge. This superheated upper layer stays suspended in the air, and can leap ridges, sail uphill and even gust across large bodies of water. It plows ahead of the heavy bottom debris which burns everything in its path.

Eruption-induced tsunami, Augustine, Alaska, USA

DURING AN ERUPTION at Alaska's Augustine volcano in 1883, a massive northern-flank debris avalanche sent a huge portion of the volcano crashing down into the sea and triggered an ocean tsunami, or tidal wave, with wave heights registered at 10m / 32.8ft. Travelling at speeds of up to 800km / 497 miles per hour, there would have been very little warning of this approaching tsunami. The first indication would have been a great sucking of the ocean out to sea, exposing the sea bed at the shore, followed by a succession of waves, the third to eighth waves usually being the biggest and most destructive. In 1986, Augustine had a moderate eruption that did not activate a tsunami, but if a volcanic eruption caused a tsunami today the populated low-lying coastline of Alaska would be in danger of serious flooding.

Volcanic lahar eruption, Nevado Del Ruiz, Columbia

THE EXPLOSIVE NEVADO DEL RUIZ volcano in the Andes volcanic chain has a history of deadly, heated mudflows. Glaciers at the summit, 5,389m / 17,680ft high, melt from the heat of eruptions, combine with volcanic ash and rocks, and surge downhill as huge, roaring lahars. On November 13, 1985, pumice and ash from Nevado showered Amero, a small town 74km / 46 miles away. Citizens were urged to stay calm but by 7pm an evacuation was ordered. Suddenly the ash stopped and the evacuation was called off. Then, at 9.08pm the volcano unleashed a violent eruption of lava and tephra melting the summit ice field, creating hot, surging pyroclastic flows and raging boulder-laden lahars that rocketted down the slope. Over 25,000 people and 15,000 animals were killed and another 12,500 people injured or made homeless.

Lateral eruption, Mt. St. Helens, USA

IN EARLY 1980 MT. ST. HELENS in the Cascade Mountain range in Washington State reawakened with a series of earth-quakes and steam blast eruptions. Two months and 10,000 earthquakes later, a visible bulge was noticed on the north side of Mt. St. Helens. On May 18, seconds after a strong magnitude 5.1 earthquake had shaken the volcano, causing a substantial debris avalanche, a huge, lateral explosive eruption blasted away the north side of the volcano. Most of its summit and flank disintegrated in a blowout of gas, ash, rocks and steam moving at 100km / 62 miles per hour for about 11km / 6.8 miles. A huge eruption column ensued, rising 15km / 9.3 miles into the sky. A cloud of millions of tons of ash blew 156km / 97 miles, settling over Spokane, Washington and in two weeks, ash from the blast had circled the Earth.

Largest eruption of the 20th century, Novarupta, Alaska, USA

THE LARGEST VOLCANIC ERUPTION in the 20th century occurred on June 6, 1912, at Novarupta volcano, one of a cluster of volcanoes in the Katmai region of Alaska causing a 4.5 x 3km- / 2.8 mile x 1.8 mile-wide crater. The blast was estimated to have the power of ten Mt. St. Helens blasts, and left behind the famous Valley of Ten Thousand Smokes. The sulphurous eruption ash cloud zig-zagged with bolts of lightning, and climbed 13km / 8 miles into the sky over Alaska, ultimately reaching Algeria. People living in the Kodiak region 161km / 100 miles away from the blast, were shrouded in darkness and developed near-suffocating respiratory problems. Roofs of homes collapsed under the weight of the ash, and communications, business and transport were all brought to a standstill.

Super-volcano, Yellowstone Caldera, USA

A SUPER-VOLCANO IS THE POPULAR TERM for a large caldera volcano that has the potential to kill millions of people, wipe out entire continents and change the planet's climate. There are only about forty super-volcanoes on Earth, and Yellowstone Caldera volcano, in the tranquil pine-forested slopes and hot springs of Yellowstone National Park, Wyoming, is one of the most dangerous. At least three ancient super-eruptions have occurred here at intervals of about 600,000 years. It has now been 640,000 years since the last Big Bang. If or, some might say, when Yellowstone erupts it will have the force of one thousand Hiroshima atomic bombs, or eight thousand times that of Mt. St. Helens' 1980 eruption, causing a global disaster, never witnessed before by modern man.

Most catastrophic eruption, Krakatau

KRAKATAU VOLCANO STRADDLES the Sundra Strait between Indonesia and Sumatra. Its devastating eruption in 1883 had the loudest documented blast – heard as far as Perth in Australia, 3,100km / 1,926 miles away. Pyroclastic surges from the explosion raced 40km / 24.8 miles to the coast of Sumatra, the resultant tsunami causing rampant devastation, and 36,000 plus fatalities, which combined to make this eruption the most catastrophic of modern times. The force of the 40m- / 131.2ft-high tsunami waves smashing into Sumatra's coast carried a large steam ship 2.5km / 1.5 miles inland, leaving it perched on a hilltop and the ash plume at Krakatau's caldera is said to have been 100km / 80 miles high. Krakatau is still active, and had its last eruption in 1995.

Most fatalities from an eruption, Tambora, Indonesia

TAMBORA, A MASSIVE STRATO-VOLCANO on Sumbawa island in Indonesia, had an enormous volcanic eruption in April 1816, and is credited with the highest number of fatalities – over 80,000. The explosion severed 1,335m / 4,380ft off the top of Tambora's summit. So much volcanic ash was tossed into the atmosphere that global temperatures lowered worldwide, and 1816 became known as the 'year without a summer'. It snowed in June in New England, USA, with the cold temperatures in both Europe and North America causing crop failure, which is estimated to have caused an extra 100,000 deaths, through starvation.

Europe's deadliest volcano, Vesuvius, Italy

VESUVIUS, RULING OVER THE BAY OF NAPLES in Italy, is one of the world's most famous volcanoes. Its cone actually grew inside the historic Monte Somma volcano about 17,000 years ago. Vesuvius is capable of unleashing deadly pyroclastic flows like the one that buried the ancient city of Pompeii in 79AD. Today, close to four million people happily live in Naples and with over 20,000 people per 1sq km / 0.3sq miles, it is the most densely populated metropolis in Europe. Like the ancient citizens of Pompeii, today's residents have forgotten that Mt. Vesuvius is a killer volcano that has been erupting for over 34,000 years. In geological terms, it is overdue another deadly eruption and when Vesuvius does erupt in the future, that eruption will be cataclysmic.

Europe's tallest volcano, Etna, Sicily, Italy

MT. ETNA RISES MAJESTICALLY 3,350m / 10,990ft above the island of Sicily. Those tenacious enough to make the demanding four-day climb to the summit are rewarded by a horseshoe-shaped caldera, the Valle del Bove. Etna's great height encompasses several climatic zones, with Mediterranean palm trees and cactuses near the base. Higher up, the climate is temperate with fruit and chestnut trees. Alpine firs grow on the lower summit but, above 1,900m / 6,233ft, only a few hardy plants survive, dotted about the vast landscape of hardened lava flows. The summit itself can be covered in snow well into summer. Etna's eruption records, dating to 1500BC, show both effusive and explosive past eruptions.

Most successful eruption evacuation, Rabaul, Papua New Guinea

IN 1994, ON THE GAZELLE PENINSULA on the island of New Britain, Papua New Guinea, elder members of the community who had lived through, and survived, an eruption in 1937, noticed the volcano was once again showing signs of coming back to life. They noticed the ground shaking, birds abandoning their nests at the base of the volcano, and dogs sniffing and barking at the ground, all of which caused residents to hold talks and draw up a plan for evacuation. Hazardous zones were clearly marked, and routes to safety zones with supplies were created. Practice evacuations were held and, when signs of an eruption increased, 30,000 people were safely evacuated. Just as the last few people were being evacuated, Rabaul did indeed erupt, destroying 75 per cent of the homes that had been occupied the day before.

▽ Effusive eruption, Kilauea, Hawaii, USA

KILAUEA VOLCANO IN HAWAII is the most active volcano on Earth with over 80 eruptions in the last 200 years. Kilauea is known worldwide for its effusive eruptions, though some are explosive. Lava can be thick (viscous) as cement, or runny as syrup depending on its composition and gas content. Generally, the more gas it contains, the denser and more explosive the magma is. Kilauea has thin, very fluid (low gas) effusive basalt lava flows that stream and travel across the landscape quickly. Effusive lava flows are chiefly associated with shield volcanoes, like those in the Hawaiian islands. Over time, thousands of layers of effusive lava flows pile on top of each other (like cake layers), forming gently sloping shield volcanoes, as happened here at Kilauea.

New island volcano, Kavachi, Solomon Islands ▷

THE SOLOMON ISLANDS include a shallow, submerged, new island volcano, smoking and burning above the waves of the South Pacific Ocean. Kavachi is a very lively submarine volcano, and sits just 30km / 18.6 miles from the subduction zone of the Indo-Australian plate under the Pacific Plate. Kavachi's first documented eruption was in 1939 when islanders reported seeing fire coming out of the water. Underwater eruptions and eruptions just on the surface create small islands, and eject steam, ash, lava flows and red-hot lava bombs. In fact nine of Kavachi's eruptions have created temporary new islands that are subsequently eroded by the waves. Kavachi also ejects plumes of mud and hot ash 3.5km / 2.2 miles above the ocean.

40

THE BIG BANG

◁ Explosive eruption, Santa María, Guatemala

GUATEMALA IS HOME TO THE EXPLOSIVE, pointed Santa María volcano that nestles in the forested highlands near the market city of Quetzaltenango. Santa María has very thick, viscous andesitic lava with a high silica content that is very explosive. In 1902 a massive Plinian eruption, the second largest in the twentieth century and lasting nineteen days, blew a 1.5km- / 0.9 mile-wide crater on the south-west flank, destroying a huge portion of Guatemala. The eruption plume was 28km / 17.4 miles high, and ash reached as far as California. Since then a huge explosive dacitic lava dome, christened Santiaguito, has been building in the crater and frequent pyroclastic explosions, lahars, and explosions have occurred.

Fumarolic eruptions, White Island, New Zealand

LOCAL MAORI PEOPLE CALL THE ACTIVE volcano in New Zealand's Taupo Volcanic Zone 'Whakaari'. Eruptions featured in several legends, created by the early Maoris that describe a fire demon rising from the water. Today a gaseous white cloud is frequently seen rising above White Island's big crater as White Island volcano has frequent emissions of gas and steam, and continuous fumarolic eruptions from several large vents that make loud hissing and squealing noises. Donald Duck and Noisy Nellie are two such vents that emit up to several thousand tons of hot gas, up to 800°C / 1,470°F, into the atmosphere each day. Predominantly steam, carbon dioxide and sulphur dioxide, these acidic gasses combine with rainwater to produce an acid fog that can affect breathing, burn skin and eyes, and kill vegetation.

Decade Volcano, Etna, Sicily, Italy

MT. ETNA IS ONE OF 16 DESIGNATED Decade Volcanoes, as named by the International Association of Volcanology and Chemistry of the Earth's Interior (IAVCEI). To earn Decade Volcano status, a volcano must be historically dangerous and sit close to densely populated areas and, in some cases, be overdue by geologic standards for a big eruption. The IAV-CEI aim to educate people living on or near dangerous volcanoes how best to survive a volcanic disaster. The European Union has provided much-needed funding for studying Decade Volcanoes in Europe, and in 1992 the Decade Volcano Programme helped divert a lava flow from Etna that was threatening the town of Zafferana. Giant concrete blocks were placed in front of the flow, diverting it away from the town.

Volcanic hazards, Etna, Sicily, Italy

MT. ETNA HAS THE POWER, potential and history to be a deadly, violent and destructive volcano, particularly if a large summit eruption and collapse occur, as geological evidence shows has happened in its eruptive past. It is located in a highly populated area, towering above the Sicilian city of Catania, and while Etna has more effusive eruptions than explosive ones, it has had episodic explosive flank eruptions as recently as 2003. The deadly hazards include lava reaching a temperature of 1,150°C / 2,100°F that easily burns vegetation, buildings and homes, flying lava bombs, and poison gasses including hydrogen sulphide.

Geyser eruptions, Pohutu Geyser, New Zealand

THE BEAUTIFUL POHUTU (the Maori word for 'big splash') Geyser sits in the middle of Geyser Flat in the spectacular geothermal Whakarewarewa region of the North Island of New Zealand. Geysers are boiling, fountaining, hot springs that gush frothy water, mineral matter, mud and steam at regular intervals. Water from snow, rain and streams slowly percolates thousands of feet down through cracks and crevices, and is then heated by an underground volcanic heat source. The hot water becomes pressurized and then shoots up to the surface, its power increased by geyserite, a substance made mostly of silicon dioxide, that tightly seals the geyser conduit as the water blasts to the surface. A narrowing at the neck or exit point forces the blast even higher. With less than 700 geysers in the world, they are a rare geological treat.

▽ Mud pots, Rotorua, New Zealand and Namafjall, Iceland ◁

GEOTHERMAL MUD POOLS, sometimes called 'mud pots', are heated by hot subterranean rocks, and burp and pop out gassy bubbles. Both Iceland and New Zealand are famous for their bubbling mud pools that exist in areas with scant ground water and extremely hot geothermal activity, featuring rising steam and acidic gas. What little water there is rises, bubbling and sizzling, through layers of grey volcanic ash and acid clay. The pale grey simmering mud is thick and soupy; if it rains it becomes thinner. Loud plops, pops, and splats are heard as gas bursts and liquified mud splatters over the rim of the ponds. If the mud is streaked with yellow sulphur or red iron deposits, the mud pot is sometimes called a 'paint pot'.

Rift Zone eruption, volcanic vog, East Rift Zone Kilauea, Hawaii, USA

A RIFT ZONE IS A VOLCANIC REGION undercut by volcanic dykes that's pocked with cracks, faults, and vents. Kilauea's East Rift Zone erupts more than 500,000cu m / 17,657,333cu ft of lava daily, which flows miles down through a complex underground plumbing system of lava tubes into the Pacific Ocean. Puu Oo vent, on the East Rift Zone, burps about 1,000 tons of sulphur dioxide into Hawaii's atmosphere each day, sometimes creating a potentially hazardous volcanic smog called 'vog'. More than 90 per cent of Kilauea is covered with lava less than 1,100 years old – in geological terms this volcano is still incredibly young. The current East Rift Zone eruption began in 1983 and shows no sign of stopping anytime soon.

Most southerly active volcano, Mt. Erebus, Ross Island, Antarctica

THE MOST SOUTHERLY ACTIVE VOLCANO on Earth is Mt. Erebus on Ross island, in Antarctica, just 1,387km / 861.8 miles from the South Pole. It has been continuously erupting since 1972, and there's an active lava lake simmering near its summit in a classic battle between fire and ice. Captain James Ross discovered the glacier-covered volcano in 1841, and the summit was first ascended in 1908 by Ernest Shackleton's exploring party. Today the McMurdo Scientific Research station is located near the volcano. One reason Mt. Erebus is particularly interesting is because the bottom half is a shield and the top a strato-volcanic cone. Generally it is called strato-volcanic and many of its eruptions are Strombolian. In Greek mythology, Erebus is the son of the god of chaos.

Steam eruption, Arenal, Costa Rica

ARENAL VOLCANO WITH ITS STEAMING, dark cone towering over the tourist town of Fortuna, Costa Rica, is a continuing threat to neighbouring towns and villages. Arenal is composed primarily of andesite and is considered to be a very young volcano in geological terms, being less than 3,000 years old. In the first part of the twentieth century Arenal was thought to be a pretty, forested hill but it burst to life with a series of earthquakes and a deadly, explosive eruption in 1968. Several burning avalanches, following the channel of the Tabacon River, obliterated the villages of Pueblo Nuevo and Tabacon killing 87 people. Today it still expels explosive, steam eruptions from the crater lake and the area surrounding the volcano – home to many hotels, complexes and restaurants and Tabacon hot springs – is now designated a high-risk area.

Lava dome eruption, Soufrière Hills, Montserrat

VOLCANIC LAVA DOMES, or plug domes, are humped mounds of very stiff, viscous lava composed of dacite and rhyolite. These high-silica lavas do not flow like basalt but plug the vent: the surface hardening while new magma bulges up underneath. Sometimes the sheer weight of the dome makes it collapse, sending glowing pyroclastic fragments and hot gas down the volcano's flanks. Lava domes can reach several hundred meters high and build up for years before collapsing into pyroclastic flows or detonating in an explosive blast. Soufrière Hills' lava dome, on the island of Monsterrat in the Caribbean, has collapsed and exploded several times since 1995 sending pyroclastic flows down the White River Valley into the surrounding sea, forcing thousands of people to abandon their farms and homes.

Volcanic ash fall hazard, Pinatubo, Philippines

ERUPTED VOLCANIC ASH consists of coarse, caustic, jagged, tiny pieces than can be smaller than 0.001mm / 0.00004in – of blasted glass, minerals and rock that do not dissolve in water. When soaked by rain, volcanic ash conducts electricity creating electrocution hazards and power outages. Ash fall like that at Pinatubo in the eruption of 1991, turned daylight into complete darkness, creating total chaos. An explosive-type eruption expels ash into the atmosphere and the wind whips away the jagged particles, sometimes carrying them thousands of kilometers before they are deposited again. The smaller the ash grain, the farther it can be blown. Once it falls and settles, it becomes as thick as cement and incredibly heavy. It kills vegetation, increases run-off, causes breathing problems and it's weight can make roofs cave in.

Emerging volcano, Paricutin, Mexico

IN FEBRUARY 1943 the Earth shook and townspeople heard thunderous roars and booms coming from the ground near a farmer's corn field in Michoacan, Mexico. A massive crack, or fissure, ripped open the full-length of the field and the air filled with a sickening smell of sulphur as black smoke belched from the crack. Next, lava, blocks and ash violently sprayed out while amazed spectators watched the birth of a brand new volcano, Paricutin. That initial eruption lasted nine years, with lava flows covering 64sq km / 24.7sq miles, and destroying villages, livestock and farms. When the eruption ceased in 1952 a 410m- / 1,345ft-high round, ashy cinder cone stood in the cornfield, hissing black smoke and it is still actively spitting gas and steam.

◁ Volcanic steam rings, Etna, Sicily, Italy ▷

VOLCANIC STEAM AND GAS RINGS are rare and beautiful. These amazing three-dimensional hoops can be 200m / 656ft wide. Visitors marvel at how the volcanic steam rings are formed, but scientists believe they contain a lot of water vapour, and are possibly spouted by pulsing steam blasts through narrow, cylindrical, volcanic vents. As recently as 2006 the Bocca Nuova vent at Mt. Etna, in Sicily, has popped out some beautiful steam rings to the amazement of watching geologists, giving them plenty to discuss.

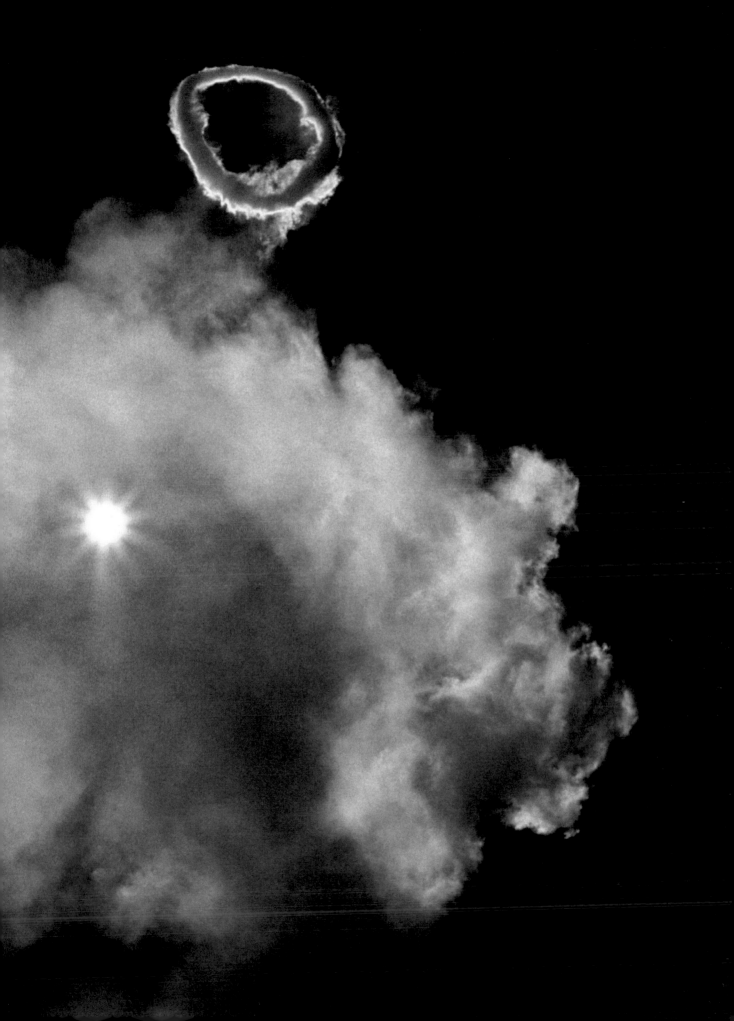

Gas eruptions, Guagua Pichincha, Ecuador ▷

MANY SCIENTISTS BELIEVE a good portion of the atmosphere that we depend on to live was boiled out of the Earth's molten interior and emitted through volcanoes. The oceans too may have formed from condensed water in volcanic eruptions. In this image steam and gasses are seen escaping from the strato-volcano Guagua Pichincha, forming a smooth lenticular cloud. Scientists are currently investigating how a basic volcanic gas, carbonyl sulfide, common in volcanic eruptions, combines with amino acids to form peptides, which are necessary for living cells and may have helped life evolve on planet Earth.

Sulfur eruptions, Vulcano, Italy

THE SOUTHERNMOST ISLAND in the Aeolian chain in Italy is the volcanic Vulcano island in the Tyrrhenian Sea. You can smell Vulcano before you arrive because the large quantities of sulphur produced by the volcano permeate the air with an acidic, rotten-egg smell. The volcanic gasses erupt from fumaroles, and hydrogen sulphide in the gas cools and condenses around the openings of the vents into pure yellow sulphur. In fact from Roman times to the 1800s the principle economic activity on Vulcano was harvesting sulphur. Today most locals make their income from tourists who come to the picturesque volcanic island from all over the world, partly to visit its hot springs and sulphur mud baths.

PYROCLASTIC CONES (FROM THE GREEK 'pyros' meaning fire and 'klastos' meaning 'broken) are created when volcanic fragments, or ejecta, are erupted up and out, then fall back around the vent from which they came. They can be simple cones built around a central vent or they can form part of a complex volcanic structure which has a dome and an erupting vent or multiple vents. The composition of the erupting molten magma rock determines three things: the type of lava, the type of eruption and the shape, type and profile of the resulting volcanic cone – spatter, tuff, cinder, ash, pumice or something other. For example, the viscosity of lava determines its flow rate – how far it will travel from the vent before cooling. This in turn influences the gradient of the slopes – how steep the sides of a volcano will be as the layers of lava build up. Cones can be just a few feet tall or, like Mauna Loa in Hawaii, soar 9 km / 5.5 miles from their base.

Volcanic cones have three basic profiles: shields, like Fernandina volcano in the Galápagos Islands, are gently sloping cones, formed by fluid lava. Composite cones, found at strato-volcanoes are sharply pointed and composed of layers of ash and lava. Cinder cones such as at Paricutin volcano, Mexico, are rounded mounds of cinders with an indented depression on top. Of course where there is a volcanic cone, there is likely to be (or have once been) a vent and a magma chamber, and craters and calderas are formed from an explosive eruption or a collapse occurring due to a sudden draining of magma from this underground chamber.

Any such depression smaller than 1.6 km / 1 mile is a crater, and any larger than 1.6km / 1 mile is a caldera (meaning 'cauldron'). Some calderas are enormous, measuring more than 50 km / 31 miles and up to

100 km / 62 miles across and over a kilometer deep. Such vast chasms can often be a sign of a super-volcano having erupted in the past. And when the magma chamber underneath these giant calderas refills with magma from inside the Earth the floor of the caldera literally rises up (and is termed 'resurgent'), possibly indicating a future catastrophe. Resurgent calderas are some of the largest volcanic structures on earth.

Sometimes craters or calderas contain evidence of past eruptions, such as the jutting nose of an ancient lava flow, protruding above the surface of a lake. These formations can often occur in thick, non-fluid lavas (andesite and rhyolite), when the hardened lava within the central conduit of a vent remains standing long after the rocks around it have eroded away. When gasses contained within rising magma become trapped under the caldera floor , the crater can also build a plug or dome which becomes highly pressurized then explodes in an big blast. And so, begins the cycle of volcanism once again.

CONES, CRATERS & CALDERAS

2

Perfect volcanic cone, Mt. Fuji, Japan

VOLCANOES COME IN ALL SIZES AND SHAPES, and have a central vent, an opening in the Earth's crust, where volcanic matter erupts. As pyroclastic fragments, lava, cinders, ash and pumice are ejected, they build up around the vent forming volcanic cones. Depending on the composition and size of the ejected fragments, different forms of cones are created. Classic strato-volcanoes like Mt. Fuji tend to have sharp-pointed cones, its flawless, symmetrical cone is made of layers of lava and ash that build up close up to the central vent. Mt. Fuji has had about 15 historic eruptions, and its graceful profile has made it one of the world's most famous volcanoes.

Previous pages | Tengger caldera, Indonesia

THE TENGGER CALDERA is a large 10 kilometer- / 6 mile-wide, crater-like structure that is the remnant of a volcano that exploded, blowing off its top and forming this distinctive landscape. The caldera floor is covered with a plain of ash, the Tengger Sand Sea.

Spatter cone, Ol Doinyo Lengai, Tanzania

WHEN MOLTEN LAVA THE CONSISTENCY OF TOFFEE, is shot out of a vent, irregular blobs break up and fly apart before splattering back to the ground around the vent in mounds. The partly molten spatter-clots dribble down the sides of the mound, cool and meld together in a lumpy, bumpy, cone shape. Spatter cones form at the summit or slope vents of bigger volcanoes that produce very fluid lavas. Ol Doinyo Lengai volcano in Tanzania is one of the only volcanoes in the world that produces an extremely fluid natro-carbonatite lava, that has a very low viscosity lava that cools very quickly with a dusting of white sodium carbonate. This is due to a chemical reaction with atmospheric water.

61

Ash cone, Meru, Tanzania

AFRICA'S MERU VOLCANO IS AN ACTIVE VOLCANO WITH A GRITTY ASH CONE around its central vent. Smooth, dark ash cones such as this form when a mix of gas and steam expands, bursts and blows the lava into tiny bits, about the size of a grain of sand or as big as a piece of rice. If the ash stays loose and gritty and doesn't compress and harden, an ash cone forms. Meru's ash cone sits inside it's breached U-shaped caldera that formed about 7,000 years ago. The ash cone has erupted before, and is capable of erupting again.

'SP' cinder cone, Arizona, USA

CINDER CONES ARE CREATED WHEN LOOSE, uneven volcanic lumps the size of irregular golf balls build up around a vent. As gasses propel magma to the surface vent, pressure builds and blasts the lava into craggy fragments at the moment of eruption. These chunks cool in flight as they shoot through the air, fall and create a steep, dry, loose cinder cone. These can erode quickly but in calm, dry climates, as in south-west USA, they last longer. The SP crater in Arizona has a perfectly symmetrical cinder cone, named the 'Shit Pot' by early explorers because of its shape. Most cinder cones are less than 33m / 108ft high and are built in one eruption. Most lava fountains form cinder cones.

Cinder cone volcano, Cerro Negro, Nicaragua

A CINDER CONE THAT HAS SUCCESSIVE eruptions, is called a cinder cone volcano. Cerro Negro (meaning 'black hill') is a good example, being about 250m / 820ft high in the Marrabios range. It is the youngest volcano in Central America and Nicaragua's most active one, and it consists of basalt cinders. It has erupted twenty times since 1850, each explosive blast adding a new layer to the cinder cone base. Volcanic cinders are sometimes called lapilli, an Italian word meaning 'little stones'.

◁ Tuff cone, Darwin, Galápagos Islands △

TUFF CONES, FORGED BY BASALT magma-water (or 'phreato-magmatic') eruptions, are rare, and are usually found near a coastline or just off-shore. Their formation demands a constant supply of water that's in contact with hot, rising magma that explodes underground. This blows tiny fragments of volcanic glass and rock up, out and down, around the vent, creating a hat shape with sharply inclined sides, at more than 25 degrees. Tuff cones can be as wide as 1.5km / 0.9 miles and 300m / 984ft high, and usually contain volcanic glass in the mineral mix. The explorer Charles Darwin first saw tuff cones in the Galápagos Islands and theorized correctly how they were formed.

◁ Eroded tuff cone, Diamond Head, Hawaii, USA ▽

DIAMOND HEAD, AN ERODED TUFF CONE, sits on the shore of Waikiki Beach in Honolulu, where native Hawaiians call it 'Leahi'. The embedded, glittering calcite crystals sprinkled in its flanks tricked British explorers in the 1800s into thinking they were diamonds, hence the name. Diamond Head formed about 200,000 years ago and was probably blasted open in a day when the Pacific Ocean entered the vent underwater, hit hot magma and blew straight up, leaving the tuff cone's symmetrical steep sides and memorable silhouette. Thought to be extinct, Diamond Head is Hawaii's state monument, and one of its biggest tourist attractions.

Steaming cone, Colima, Mexico

MANY VOLCANOES ERUPT STEAM as well as ash, rocks and lava and Volcan de Colima in Mexico at 4,100m / 2.5 miles high, with its classic conical profile, regularly spews steam from its pointed cone such as in 1999 when a white eruption cloud of water vapour and sulphur towered 5km / 3.1 miles above the 'Volcano of Fire'. Steam eruptions can be violently explosive, occurring as hot, surging magma up to 2,000°C / 3632°F encounters ground or surface water, and instantly turns to hissing steam. In 1949 Thomas Jaggar termed this a 'steam-blast eruption'. Colima is Mexico's most dangerous volcano because of its frequent explosive eruptions, and as 300,000 people live nearby.

Scoria cone, Komatsuka cone, Aso, Japan

SCORIA IS A REDDISH-BROWN IGNEOUS ROCK (or pyroclast), with 6.4cm / 2.5in fragments, bigger than cinders and glassier than pumice. It forms when pressurized basalt magma that is saturated with trapped gas bubbles cools, creating thousands of small, hollow, sphere-shaped voids. Entire volcanic cones of scoria have been created at Komatsuka Cone on Aso volcano, in Kyushu, Japan. Thread-lace scoria, also called reticulite, is a very rare form, which is created when a thin, frothy layer of gasses, similar in consistency to foam on top of a mug of beer, bursts open on top of basalt lava flows, and cools in the air. Thread lace scoria is 99 per cent porous, and is the lightest rock on Earth.

69

Pumice cone, Tullu Moje, Ethiopia

PUMICE IS PALER THAN OTHER PYROCLASTS, being glass-rich with a high-silica content, and dotted with holes like a sponge. Lightweight and pocked, pumice rocks are 90 per cent porous and will float on water. They are made when frothy, gassy, viscous lava, such as rhyolite, is tossed into the air and cools quickly. Tullu Moje, in Ethiopia, is a young cone composed of erupted pumice. The ancient Romans used pumice to make a lightweight, concrete building material, and it is also used to grind off callous skin.

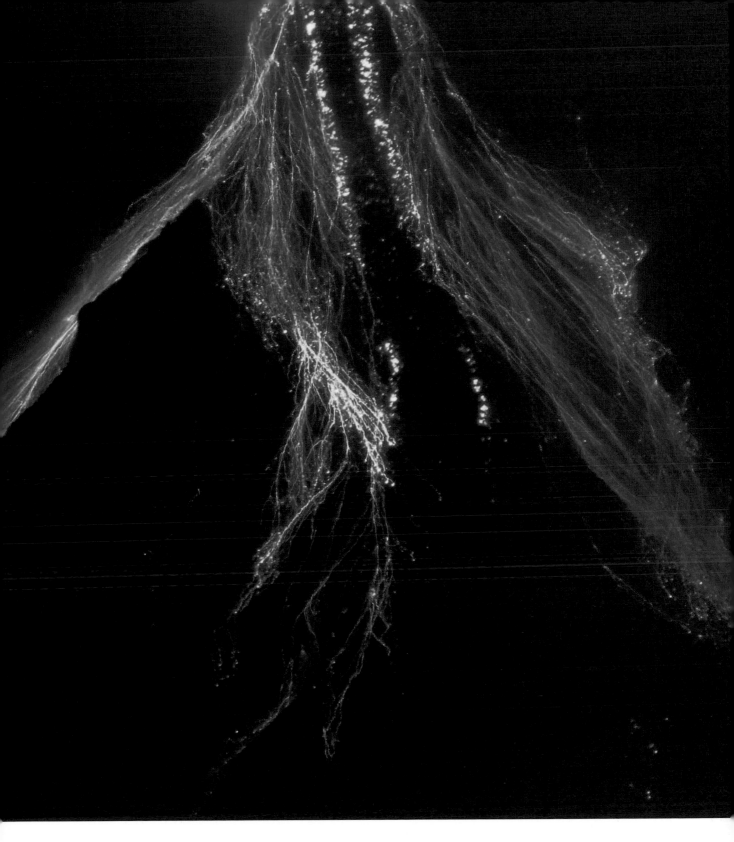

Pyroclastic cone, Mayon, Philippines

VIRTUALLY ALL VOLCANIC CONES were originally pyroclastic. This simply means they are built of fragments of fire-hewn rock and bits of cooled lava. Pyroclastic cones can refer to the small 'parasitic' kind, located on larger volcanoes, or to giant volcanic structures. Mayon volcano, in the Philippines, has a perfectly shaped pyroclastic cone, and straddles a convergent plate boundary. At this point, located on the Pacific Ring of Fire (which encircles the basin of the Pacific Ocean), the Eurasian Plate overcomes the Philippine Plate, forcing the rock down into the Earth's mantle, where it melts and forms magma and gasses, bubbling under the Earth's crust.

Chain of cones, Cascade range, Oregon, USA

VOLCANIC CONES CAN SIT ALONE or form long chains as in the state of Oregon, which is criss-crossed with a 50km- / 31 mile-wide band of cinder cones and strato-volcanic cones. The state's Western Cascades are a chain of older, worn-down volcanoes that formed 40 million years ago along the boundaries where the North American and Juan de Fuca Plates converge. Younger regions in the Cascades are called the High Cascades. From British Columbia to Northern California, 13 active volcanic hubs lie in an arc-shaped strip along the coast like a necklace of explosives. Many people living in the Pacific Northwest region of the USA are not aware that they are surrounded by the potentially deadly volcanism of Mt. Shasta, Mt. Ranier, Mt. Bachelor, Mt. Hood and the other Cascade volcanoes.

Explosive Andesitic cone, Lonquimay, Chile

LONQUIMAY IS A SQUARISH-TOPPED VOLCANO, with an explosive andesitic cone, which nestles deep in the Andes Mountains of Chile. A sprawling glacier plunges down the southern slope from the summit, and the entire volcano is slashed in half by a long fissure known for its myriad of lava flows. In fact the majority of the Earth's land volcanoes are andesitic. Andesite lava is pale and dense because it contains more than 50 per cent glassy silica, and encompasses explosive, super hot gas. The thick magma builds up pressure in the vent and on top of it, sometimes creating lava dome plugs, before finally erupting in giant blasts. This Andean sleeping beauty is frequently draped in a mantle of pristine white snow that entices a few brave ice climbers to its windy summit.

Basaltic shield volcano caldera, Kilauea, Hawaii, USA

VOLCANIC CALDERAS ARE SOME of the most fantastic features on Earth. They can contain lava lakes, acid lakes, hot springs and other geothermal activity. Most shield volcanoes, including Kilauea in Hawaii, have collapse calderas, which formed in a sudden crash when magma raced away from a shallow storage chamber beneath the volcano. The collapse crater on Kilauea, formed in the mid-1980s near Puu Oo vent, once contained a splashing lava lake inside. Basalt lava flows have been continuously erupting at Kilauea since 1983, making it the longest eruption in recorded history. Since such basalt lava contains less than 55 per cent silica it can flow like watery syrup and Kilauea's lava runs 4.5km / 2.8 miles down slope, and pours into the Pacific Ocean.

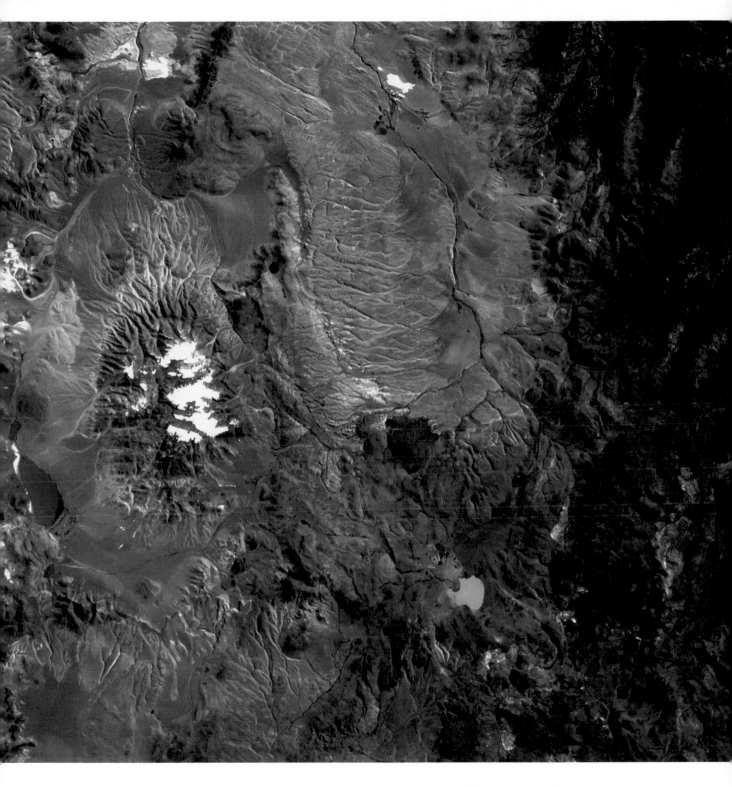

Resurgent caldera, Cerro Negro, Argentina

IN THE ANDES MOUNTAINS OF ARGENTINA, Cerro Negro's sweeping, resurgent caldera is one of the largest volcanic structures on Earth. Resurgent (literally, 'rising') calderas are linked to mammoth eruptions the likes of which we haven't experienced in modern times. They form when giant masses of viscous, gas-laden lava surges up and bursts out of the Earth, followed by vast pyroclastic flows. The caldera then collapses in on itself when all the volcanic material has erupted. The resurgence of the caldera floor happens over hundreds of years as new magma refills the volcanic chamber below. Cerro Negro was formed 2.2 million years ago and the crater is shaped like an ellipse, 35km / 21.8 miles wide.

East Rift valley caldera, Ethiopia

AFRICA'S EAST RIFT VALLEY stretches across Tanzania, Kenya and Ethiopia, and is in a highly volcanic area where three plates of the Earth's crust join. Ethiopia has more than forty volcanoes, and rivers pouring down into the country's rift form lakes in craters and on the savanna grasslands. O'a caldera is Ethiopia's largest rift caldera at 15 x 25km / 9.3 x 15.5 miles. The giant caldera is home to Lake Shala (at bottom), believed to be the deepest lake in North Africa. It is the primary breeding ground for Africa's great white pelicans.

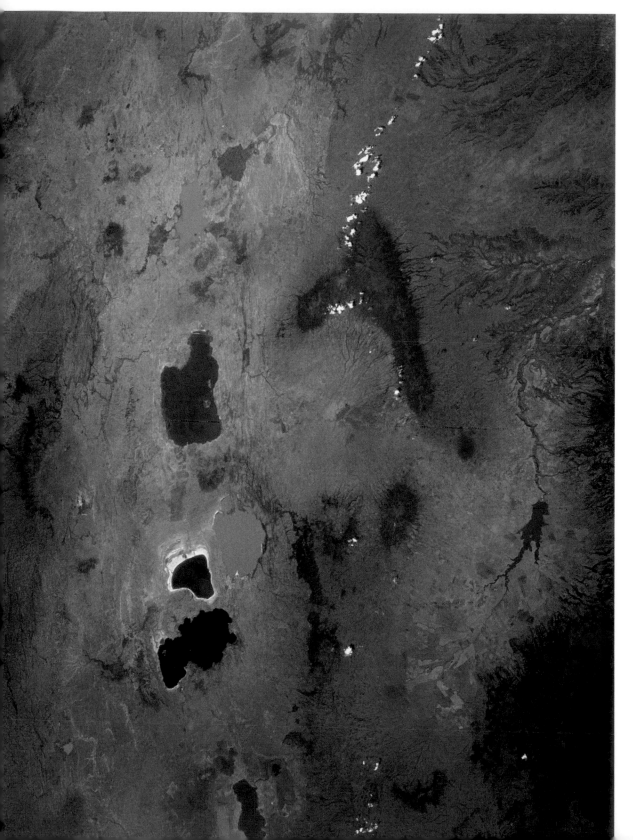

Caldera volcano, Toba, Indonesia

TOBA VOLCANO IN INDONESIA has a mammoth, treacherous caldera 100km / 62 miles wide that is home to a vast lake. The volcano's last eruption 75,000 years ago was the biggest eruption on Earth in the last two million years, and involved more than 2,800 times the amount of volcanic debris that Mt. St. Helens blasted into the atmosphere in 1980. Some scientists believe that the blast brought humans to the brink of extinction through global climate change. The region is still seismically active, with the Indian Oceanic Plate continuing to subside beneath the Eurasian Plate, supplying magma for eruptions. If Toba were to erupt today, it would be a cataclysmic super-eruption creating a global 'volcanic winter', killing crops and leading to mass starvation.

Largest caldera on earth, Yellowstone, Wyoming, USA

MANY EXPERTS BELIEVE YELLOWSTONE caldera volcano, is the most perilous volcano on Earth. The huge caldera, in Wyoming, USA, collapsed after a massive eruption about 600,000 years ago and today the earth under Yellowstone is refilling with magma – reloading – for another calamitous blast. It was declared the world's first National Park in 1871, and visitors flock to see shooting hot geysers, bubbling mud pots, colourful mineral terraces, rock canyons, wolves and bison. If Yellowstone erupts again most of North America will be covered by volcanic ash, rock and debris, bringing much of the country's infrastructure to a halt. Global cooling on a disastrous level would follow with many subsequent, devastating effects to the ecosystem. Yellowstone's destructive potential is in a class of its own.

Double caldera, Telica, Nicaragua

TELICA VOLCANO, ONE OF THE FEISTIEST in Nicaragua, has a double summit caldera formed by two different collapse episodes. One crater is older and shallower on the southwestern slope. The second, a 120m- / 394ft-deep young crater on the south summit, is alive with active fumaroles, steam emissions and recent eruptions. The slopes of Telica are barren from much ash fall. Telica is just one of a group of volcanoes – the Telica Group – whose eruptions date at least as far back as the sixteenth century. In 1994 Telica shot a gas and ash plume 800m / 2,625ft over the summit, with ash falling 25km / 15.5 miles away. Nearby is Hervideros de San Jacinta, a popular tourists spot bubbling with mud pots.

Inhabited crater, Pululagua, Ecuador

PULULAGUA (ALSO KNOW AS PULULAHUA) is a bustling town and agricultural mecca in the bottom of a huge volcanic crater, just north of Quito, Ecuador. (Pululagua means 'smoke of water' to the indigenous Quichua people.) The first to settle it were ancient Incas, followed by the Dominican Monks in the 1800s and the modern agricultural town of Nublin now prospers inside the large crater because of the rich volcanic soil. In 1969 Pululagua was made into a National Park, and is now home to various unique botanical species and animals including bats, lizards and wolves.

Float-in, flooded caldera, Deception Island, South Shetland Islands

DECEPTION ISLAND IS AN INCREDIBLE, active volcano in the Shetland Islands, near Antarctica. Frozen glaciers covering half the volcanic island are streaked with dusky ash. One wall of the flooded Foster caldera has been breached, allowing the sea to wash inside at Neptune's Bellows, a small, open channel. Consequently, Deception is the world's only float-in volcano allowing ships to pass into the ominous caldera. This unusual, harsh, geothermal environment is home to over a dozen species of moss and lichen found nowhere else. The south-west coast is inhabited by over 100,000 chin strap penguins. Also of note, is that the floor of the caldera is rising, indicating possible future volcanic activity.

Tengger caldera, Bromo, Java, Indonesia

TENGGER CALDERA, 16KM / 10 MILES WIDE, is part of a volcanic complex encompassing five strato-volcanoes. The youngest and most active is Bromo, a popular tourist trek in Java whose pyroclastic cone can be seen from a distance frequently belching an eruption cloud. Bromo has erupted about 53 times without warning in the last 200 years. In 2004 two tourists visiting the vent were killed by a 30-minute surprise explosive eruption of trapped gas and magma, and several others were seriously injured by flying volcanic rocks. Various myths and legends surround Bromo, and each year thousands of Indonesians trek to the lip of the crater to pay homage and make offerings of poultry, flowers and fruit, in supplication to the giant mountain.

La Cumbre caldera, Fernandina Island, Galápagos Islands

LA CUMBRE VOLCANO ON FERNANDINA ISLAND is the most active volcano in the Galápagos Islands. In the 1990s red lava oozed though cracks in the caldera floor, and in 2005 it erupted lava flows and jets of ash and steam 7km / 10.5 miles high. The volcano now rests amidst Fernandina's black lava sand beaches, while hordes of dragon-like marine iguanas dive along the rocky coast for crunchy meals of nutritious seaweed. Fernandina is a protected sanctuary where no one lives. Its population of large land iguanas however, make use of the rim of La Cumbre caldera for their nesting sites, though recent eruptions have destroyed much of their habitat.

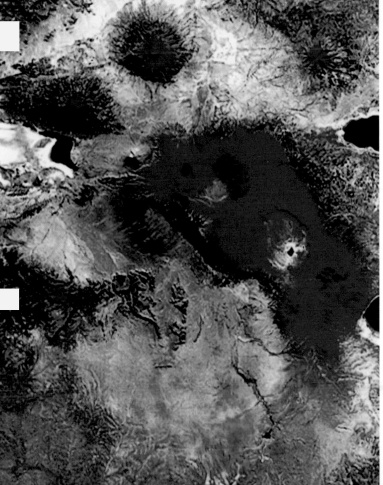

◁Ngorongoro caldera, Tanzania△

NGORONGORO CALDERA IS 22.5KM / 14 MILES WIDE and 620m / 2,034ft deep, and has a lake in the flat bottom. Wild animals, including zebras, baboons, elephants, rhinos, hippos, gazelles, buffalos, warthogs, lions and leopards drink here and graze the clover grasslands, and archaeologists have also discovered human fossils in the caldera. Today over 8,300sq km / 3,205sq miles of the caldera, which was formed when the African Rift collapsed millions of years ago, are protected, and the region is known as the Ngorongoro Conservation Area (NCA). One dirt road descends into the caldera, past native Massai villages.

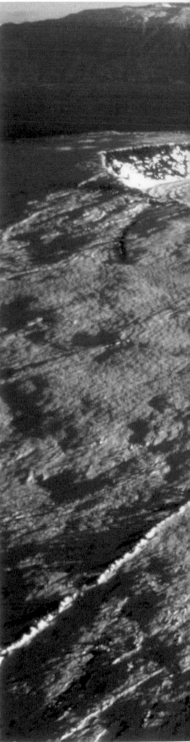

Toya caldera, Usu, Japan

MT. USU IS KNOWN FOR ITS EXPLOSIVE pyroclastic eruptions, and those from the 1600s to the present have resulted in multiple deaths from landslides and mudflows. The resort town of Toyako, at the base of the volcano, is hit
by frequent volcano-related earthquakes, while Toya Caldera is zig-zagged with dangerous faults which could trigger future pyroclastic flows and avalanches. In the year 2000 more than 1,600 earthquakes were experienced in
just one day ripping 100km- / 62 mile-wide cracks in the caldera walls, forcing tens of thousands of people to
evacuate. Toya caldera has several lava domes that pose future threats.

Multiple calderas, Mauna Loa, Hawaii, USA

MAUNA LOA'S ERUPTIONS HABITUALLY BEGIN at its main summit crater, Mokuaweoweo, as impressive lava curtains squirting from 2km- / 1.2 mile-long fissures. The earliest recorded eruption was in 1780, and there have been about 40 more since then. The largest recorded earthquake in Hawaii was a shock of magnitude 8.0 caused by the 1868 eruption of Mauna Loa. It has multiple overlapping summit craters which formed, as most shield volcano calderas do, when magma rapidly erupted from or drained away from shallow storage areas under the summit. The resulting steep-walled calderas and craters are less than 1km / 0.6 miles in diameter. The linear crater row on top of Mauna Loa may reflect the direction of the underlying fissure. Repeated summit eruptions have periodically filled the main summit crater.

Long Valley Caldera, Califonia

CALIFORNIA'S 30KM- /18.5 MILE-WIDE Long Valley Caldera borders the Sierra Nevada Mountain Range. Since the 1980s, there have been swarms of earthquakes and central uplifting inside the caldera indicating a new injection of magma below. In 1990 the news headlines reported seepage of poison CO_2 from underground magma at the nearby, trendy Mammoth Mountain ski resort. Trees died from soil laced with CO_2 and a forest ranger almost suffocated (inhaling CO_2 can cause unconsciousness and even death). Geologists believe that the massive caldera is dormant, just taking a rest, and could erupt again. Concentrations of CO_2 soil gas are now being monitored year-round at Mammoth Mountain. During the 1990s enough deadly CO_2 was leaked to fill one million wine bottles per day.

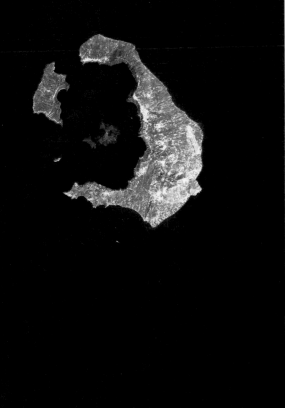

◁Santorini caldera and the legend of Atlantis ▽

WAS SANTORINI CALDERA THE LOCATION of the legendary city of Atlantis that mysteriously sunk into the Aegean Sea never to be seen again? Many believe a huge ancient eruption at Santorini sparked the Atlantis legend. The ancient Greek philosopher Plato was the first to use the name Atlantis, and he discussed a race of virtuous, divine people living on a circular island paradise who became corrupt. The angered Gods destroyed Atlantis, sinking it into the sea in a single day with earthquakes and floods. The geological evidence points to a cataclysmic eruption during the Bronze Age which demolished a Minoan culture in the Santorini Island group, and collapsed and flooded the modern day caldera. The group of islands still exhibits ongoing geothermal activity in numerous fumaroles, and hot springs.

Lakagigar, Skaftafell National Park, Iceland

LAKAGIGAR, IN ICELAND'S Skaftafell National Park, has some spectacular geological formations. The word Lakagigar means the Laki Crater Rows. These craters were formed in 1783 along a series of ten individual fissures each 2–5 km / 1–3 miles long. The initial eruption began with earthquakes, explosions and blasts of ash from the first fissure. In the eight months following nine more fissures cracked opened, erupted lava, ash and scoria and built continuous rows of craters. When the eruption finally quieted more than 125 craters had been formed at Lakagigar. The region is still considered volcanically active today.

Matupit crater, Mt. Tavurvur, Papua New Guinea

THE PRETTY TOWN OF RABAUL lives side by side with potential disaster on New Britain Island, Papua New Guinea, home to several active volcanoes including Mt. Tavurvur and its belching summit, Matupit Crater. In 1994 a blast practically buried Rabaul. When the evacuated townspeople returned, they simply dug their homes out from the ash and rebuilt the town. There are regular explosions of black, billowing ash and blasts still shatter windows, and the base of the volcano has a steaming black sand beach. No one can predict what the volcano holds in store for residents, but geologists monitor the danger closely.

Horse-shoe shaped caldera, Tungurahua, Ecuador

TUNGURAHUA, AN ANDESITIC-DACITIC strato-volcano, in the Cordillera region of the Andes, is Ecuador's most active volcano. A large debris collapse occurred here 3,000 years ago leaving behind a 183m-/ 600ft-wide horse-shoe shaped caldera. Explosive incandescent pyroclastic eruptions have occurred historically at the summit region in addition to belching steam, lava, gasses and ash eruptions. In 1999 an eruption reached the base of the volcano and forced an evacuation of the town of Banos. Although the volcano is frequently blanketed in snow locals call it 'The Black Giant'. In the local Quichua language 'tunguri' means 'throat' and 'rahua' means 'fire'.

Chinese hat crater, Galápagos Islands

THE SHIELD VOLCANOES OF THE GALÁPAGOS archipelago consist of dozens of islands, islets and rocks of volcanic origin, and some of the main islands, Fernandina and Isabela, still erupt. Like the other Galápagos volcanoes, they formed over a volcanic hot spot, a plume of magma rising from the Earth's mantle. The Galápagos hot spot rises from the sea floor and the Nazca-Cocos ridge. Small Chinese Hat, off Santiago Island, is a miniature volcano where red and yellow Sally Light Foot crabs scurry amongst the spatter cones and hardened lava rocks. Tourists come here to dive in its pristine waters in search of schools of colourful fish and manta rays.

Breached crater, White Island, New Zealand

IF EARTHQUAKES RATTLE FRACTURED CRATER WALLS they can collapse and breach, which is precisely what happened at White Island strato-volcano. New Zealand's most active volcano emerges above the waves in the Bay of Plenty, like a sea monster – just the tip of a much larger volcano. In fact 70 per cent of the giant mountain is submerged. In 1914 a rock avalanche from a crater wall collapse buried a sulphur-mining operation, killing twelve workers. Geologists speculate that acid chemicals leaching from the crater damaged and weakened the walls, causing that collapse.

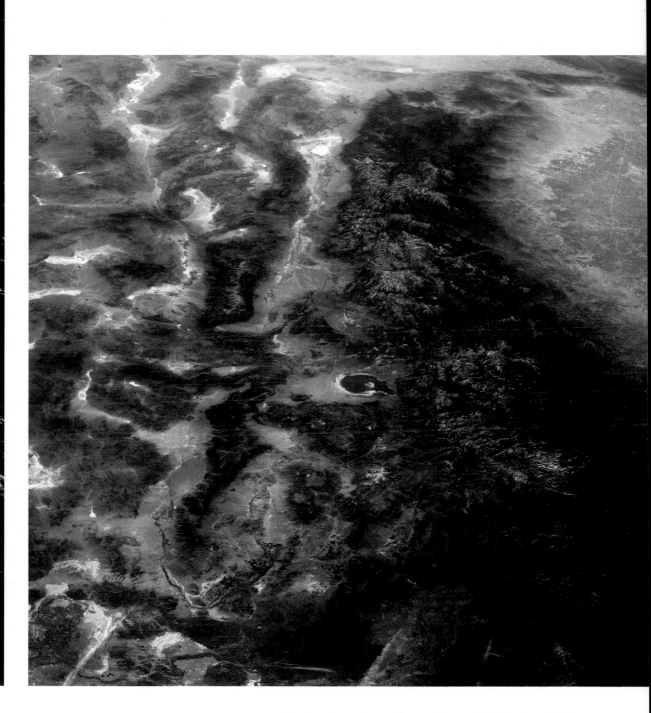

Chain of craters, Mono-Inyo, California, USA

FOR CENTURIES STEAM BLASTS, earthquakes, CO2 emissions, volcanic uplifting, explosive eruptions, pyroclastic flows and thick, inflating lava domes have been part of everyday life at the Mono-Inyo Chain of volcanic craters, a series of deep circular pits in California, USA. The active craters reach north, dissecting the Long Valley Caldera volcanic complex all the way to Mono Lake. Geologists believe that Mono-Inyo's violent history points to future eruptions. A winter eruption would be particularly devastating, melting snow and causing severe mudflows and flooding. If ash was erupted, the winds could carry it for miles before dropping it on buildings, streets and vegetation, causing massive power surges. Past eruptions at the Mono-Inyo craters have also produced fiery pyroclastic flows of hot gas and debris.

Active crater, Mt. Ngauruhoe, New Zealand

THE PERFECT, POINTED CONE of Mt. Ngauruhoe on North Island, New Zealand (part of the Tongiriro National Park volcanic complex), contains the most active crater in the country. There have been 60 eruptions since 1839, including three episodes of massive, jetting fire fountains. In 1975 the crater shot up an explosive ash plume 13km / 8 miles high. Over the decades, violent pyroclastic flows issued from the mouth of the crater and eruptions of lava have overfilled the crater, spilling down the slopes into Mangatepopo valley. Although the lively, dangerous crater is speckled with steaming, gassy fumaroles, many adventurous tourists still make the steep climb to the top. It also featured in the *Lord of the Rings* film trilogy as the dark Mt. Doom.

Snow covered crater, Mt. Ruapehu, New Zealand

IT IS ALMOST IMPOSSIBLE TO IMAGINE fiery lava shooting out of the snow-capped andesitic crater of New Zealand's Mt. Ruapehu. At 2,797m / 9,176.5ft it is the North Island's tallest volcano – and the island's most active. Melting snow and an acidic crater lake have created mudflow lahars, with waves of water, debris and ash as thick as flowing cement reaching the Whangaehu River and destroying Whakapapa's ski resort buildings. Other volcanic hazards include inflating viscous lava domes around the vent that might collapse into hot, fast, pyroclastic flows. When the volcano is quiet, people flock to the snow-covered slopes for winter sports, but a warning system has been set up to alert people of imminent danger.

Nested craters, Masaya, Nicaragua

THE SUMMIT OF NICARAGUA'S MASAYA VOLCANO is pocked with a string of pit craters, and one of the biggest is called Santiago. During the rainy season, water infiltrates the cracks turning these craters into a row of steaming kettles. Columns of 'nested' pit craters, their rims arranged inside each other, like concentric rings, are formed when underlying lava along a rift rapidly subsides and multiple collapses occur. Masaya is Nicaragua's first national park, and a road has been constructed allowing visitors to drive right up to the edge of the craters. In 2001 an explosion of rocks and debris from one of the nested craters dented tourists' vehicles in the parking area. During the time of the Spanish Conquest, in the 1500s, it was believed that gold lined the summit crater lake.

Funnel crater, Gorely, Kamchatka, Russia

THE NORTHERN KAMCHATKA PENINSULA on the Pacific Ring of Fire, is one of the most remote regions of Russia, containing over 100 volcanoes. A deep, steep funnel crater with a ragged brim sits on top of the rumbling Gorely volcano caldera complex, the fumaroles deep within sending veils of steamy vapour skywards, and the dark walls of the crater glittering with splinters of ice and pockets of snow. The Gorely complex consists of five overlapping strato-volcanoes with over 40 different craters, some containing acid lakes, and some hot springs. One of the largest earthquakes ever recorded in Russia was in Kamchatka and it registered 8.7 on the Richter scale.

Darwin crater, Tagus cove, Isabela Island, Galápagos Islands

ISABELA IS A YOUNG VOLCANIC ISLAND in the Galápagos. Here, the steep, chiselled walls of Tagus Cove are deeply carved with historical graffiti, the names and dates of pirates and whalers' ships that sailed through the narrow channel to dock in the protected cove. (Tagus is the name of a British ship that visited in 1814.) The cove's steep lava cliffs are also home to blue-footed boobies (long-winged seabirds), Galápagos penguins and flightless cormorants. A wooden stairway has been built from the cove up to Darwin Crater and the unusual, emerald-green, saltwater Darwin Lake in its centre. It's thought that this lake, above sea level and twice as salty as the ocean, was filled by a tidal wave caused by an erupting Galápagos volcano. As water evaporates from the lake, a rim of crusty white salt is left behind.

Red crater, Tongariro Massif complex, New Zealand

RED CRATER IS PART OF THE NEW ZEALAND Tongariro Massif, on North Island, a section of a volcanic mountain range with dozens of volcanic cones and craters including Kakaramea, Pihanga, Tongariro, Ruapehu and Mt. Ngauruhoe, one of New Zealand's most active volcanoes. Different eruptions have shaped the region in the last 275,000 years, but Red Crater is relatively young having formed about 3,000 years ago. Quietly resting now, it last erupted in the mid-1800s. Erosion from wind, rain and snow have exposed layers of cinders and tephra in various colors. The entire region is covered with volcanic rubble, lahars, and lava that has solidified in layers.

Ice-filled crater, Mt. Erebus, Antarctica

IN 1908, THE CREW OF THE EXPLORER ERNEST SHACKLETON were probably the first to scale Mt. Erebus in Antarctica. Their diaries describe a mighty rupture emitting bulky mounds of steam. Today, scientists monitor lava movement and eruptions at Erebus all year. Data is sent to the scientific observatory at McMurdo Sound where researchers were startled to discover the lava flows are inter-layered with ice sheets, suggesting past eruptions flowed over the glaciers without melting them. The long Erebus glacier tongue is 300m / 984ft thick in places, and fierce, gas-driven eruptions fling 8m- / 26.2ft-wide lava bombs 1000m / 3,280.8ft high in the air. Extraordinary ice sculptures also form above the hot fumaroles as the ice is melted, blown about and quickly re-frozen.

Active craters, Etna, Italy

EUROPE'S TALLEST VOLCANO, at 2,900m / 9,514.4ft high, loves to put on a show. Italy's Etna has four active craters. The hot rocks heat the ground water, converting it into steam, which hisses out of all four craters. In 2002 pulsing jets of lava leapt in surging fountains, and rivers of lava streamed down Etna from the slope, dazzling spectators. Etna's lava flows usually travel slowly enough to give people time to run but, with so many people living in such close proximity to a very active volcano, the cost of property damage is staggering. Locals in the town of Catania have nicknamed Etna the 'friendly giant'.

Erupting crater, Alaid, Kurile Islands, Russia

THE NORTHERN KURILES ARE AN ARCHING STRING of islands between the Sea of Okhotsk and the Pacific Ocean, generously sprinkled with volcanoes. The tallest is Alaid whose dark, pointed cone is layered with ash, and which occupies most of the island of Atlasova. It has blown Strombolian, Plinian and Vulcanian eruptions in the past from its central crater, as well as sideways blasts. In 1981 Alaid sent a massive ash explosion 9,000m / 29,527.5ft high.

Steaming crater, cinder cone, Gunung Bromo, Java

THE CINDER CONE CRATER OF GUNUNG BROMO, in Java, is small but quite active. It is one of a cluster of craters and cones inside the Tengger Caldera, and often spits white steam. Since 1804 Bromo has had big eruptions every few years. The indigenous Tenggerese people consider Bromo a holy mountain, home to the god Brahma and each December religious ceremonies are performed on top of Bromo after midnight during a full moon. Hundreds of tourists also make the dusty, ashy trek before dawn across the 'sea of sand', throughout the year, and climb to Bromo's crater to watch the sun rise and to witness Bromo's power, beauty and eruptions up close.

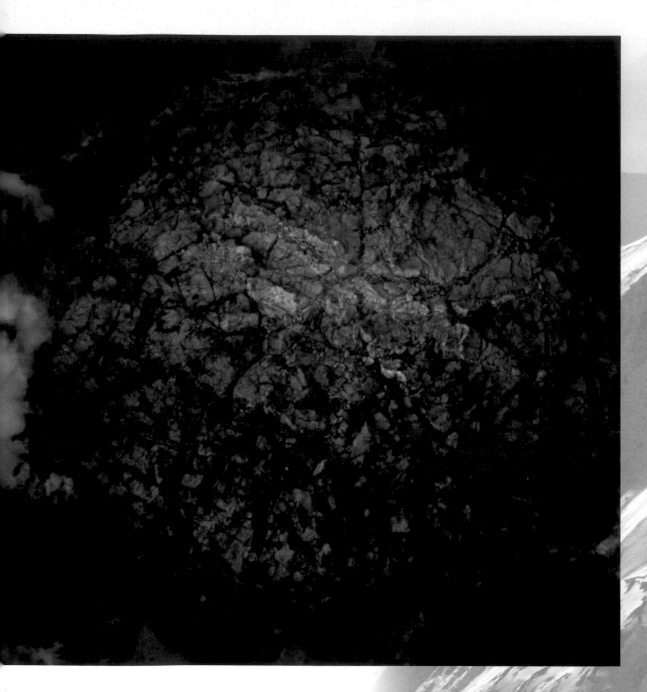

△ Growing lava dome in crater, Mt. St. Helens ▷

MORE THAN TWO DECADES AFTER its massive blast in 1980, Mt. St. Helens has a big new lava dome growing in its central vent. Because the thick, paste-like lava is too dense to flow, it squeezes out of the vent, where it sits and is inflated by hot pressurized gasses that will eventually cause an explosive eruption or a collapse into a pyroclastic flow. As the dome inflates, cracks form in the outer skin and chunks slough off. The dome has so far reached a maximum of 2,350m / 7,710ft high, and 1,030m / 3,379ft long. New lava is being added at approximately 2-3cu m / 70.6-106cu ft per second, enough to fill an Olympic-size swimming pool in about 15 minutes. Even at this incredible rate it will take centuries to replace the material blasted away in 1980.

LAVA THAT FLOWS IN RIVERS, SQUIRTS IN FOUNTAINS AND POOLS IN LAKES is quite rare compared to other erupted volcanic products like cinders, rocks, ash and gasses. Nevertheless, the visual spectacle of lava spewing out of a central vent has to be the pièce de résistance of volcanoes. From roaring lava fountains to tranquil blue lakes, this chapter examines the fluid dynamics associated with volcanoes. Moving lava will burn, bury, smother and cover everything in its path but generally flows slowly enough for people to get out of danger.

Lava can be thick and paste-like (high-silica, high-gas) or thin and runny (low-silica, low-gas) depending on the melt. Magma rises to the surface to erupt by three determining forces: bouyancy (thinner = more bouyant), gas pressure and the magma supply rate. Basalt lava is the most bouyant type of lava at an average of 2.7 grams / 1 ounce per cubic centimeter. Basalt lava also has a low gas content making it rise slowly so its gasses are released gently, resulting in beautiful oozing lava rivers or surging fountains.

The two main types of basalt flows commonly erupted at shield volcanoes are known by their Hawaiian names: pahoehoe and aa lava. Thin and runny, pahoehoe lava flows like honey and cools into a smooth or ropey sheet whereas aa lava is a little pastier, and is covered with chunks that tumble along the flow front. Basalts are some of the fastest moving flows, clocked at over 30 km / 19 miles per hour on steep slopes and these can form long rivers tens of kilometers long. Magma with higher silica and gas content (andecite, rhyolite or dacite lavas) creates less common, sluggish lavas known as 'felsic'. These build up pressure trying to rise to the vent and then explode rather then bounce up and flow. Lava can form in thick square chunks of solid lava, called blocks, which are thrown up in an eruption and are sometimes found in some aa flows.

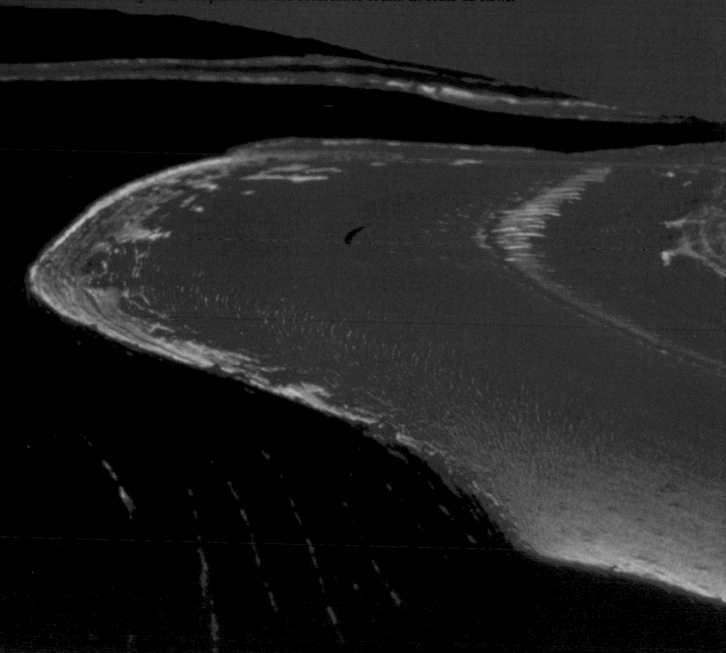

When fluid basalt lava erupts from elongated fissures a long row of lava fountains forms a stunning curtain of fire. If it erupts from a single constrained vent, gas bubbles in the melt can expand rapidly upward to form a roaring fountain that can shoot the molten rock high into the sky in weird shapes – like rooster tails.

Double fountains can also occur, shooting up from separate vents side by side. Cones will eventually be formed from such erupted material falling back down due to gravity and when the bowl shapes of craters and calderas gather rainfall fresh-water volcanic lakes are found. The water takes on extrordinary colours from dissolved minerals, acidic-chemicals and erupted gasses from deep within the earth, turning them acidic over time. These can be lethal when they overflow. Of course lava lakes (usually basalt) are simply lava filled craters, calderas on a volcano. These can be raging continually, or semi-solidifed, depending on the amount of new lava being pumped up from the magma chamber underneath.

FLOWS, FOUNTAINS & LAKES

3

△Lava flows and man, Etna, Sicily, Italy ▷

SICILY'S MAJESTIC MT. ETNA dominates the Italian island, providing fertile soil but also frequently sending lava down
its slopes. In 2001 a massive molten river of red lava threatened to overrun and destroy a tourist hotel at the base
of the volcano. Emergency labourers worked day and night to strengthen 30m- / 99ft-high barriers of bulldozed
soil placed in the path of the lava flows, and when a second river of lava advanced, but then stopped just 4km /
2.5 miles short of the city of Nicolosi, the people of Catania must have thought the old gods their ancestors feared
were alive and well. The reality is, the cessation of Etna's eruptions were more to do with the unpredictability of this
volcano than anything to do with artificial blockades.

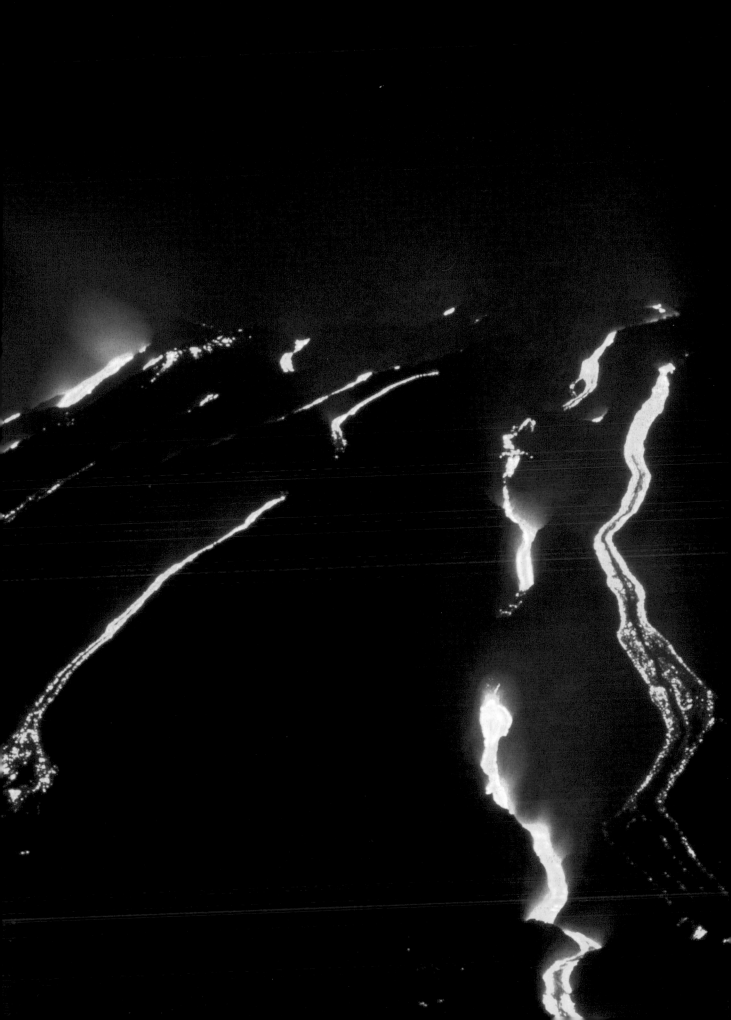

Basalt lava flows, Kilauea, Hawaii, USA

MOST LAVA IS BASALT, and basalt lava runs in streams and rivers because of the low gas and silica content. If the lava has no gas then no pressure builds up, and no explosions occur, resulting in gentle, voluminous, effusive, lava rivers and glowing lava flows generally associated with shield volcanoes. Kilauea volcano is famed for its vast basalt lava flows that quickly cover kilometers of the Hawaiian islands, flooding the land like a giant, overflowing molten bathtub. High-volume basalt lavas are relatively thin, allowing multiple flows to build up in layers over time. The lavas can vary in composition, some being quite dense, some watery. Once basalt lava cools and hardens, the compositional differences determine the texture, shapes, configurations and patterns of the cooled rocks.

Lava fire hose, Hawaii, USA

AS LAVA TUBES FEED HOT LAVA into the cold sea, hardened shelves of lava build up and hang off the coastal cliffs above them. On rare occasions a big wave or earthquake will crack the lava shelf away from the cliff face, beheading the lava tube and exposing a temporary, fabulous 'lava fire hose'. This is an arching, smooth stream of red-hot basalt lava bursting from the tube into the sea. When the tide eventually rises, the cold water seals off the tube again and the lava finds another exit through weaker cracks elsewhere.

Felsic lava, Mt. Unzen, Japan

THE MORE SILICA THE MAGMA CONTAINS, the thicker and denser the lava is. Felsic lavas are thick, viscous dacitic and rhyolitic lavas, that are as dense as concrete and are the unpredictable force behind explosive volcanoes, generally strato-volcanoes, such as Mt. Unzen. Pressure inside the volcano builds and builds because gas can't escape from the thick magma
and, when the magma reaches the vent, all the bubbles burst causing either a mighty explosion, or the formation of a big
lava plug, up to 500m / 1,640ft thick, that eventually collapses or detonates. Scientists study violent eruptions in the hope
of finding new information to prevent the kind of deadly disaster that happened at Mt. Unzen in 1991, when 41 people,
including the French volcanologists Maurice and Katia Krafft, were engulfed and killed by an unexpectedly violent blast.

Andesitic lava, Sakura-jima, Japan

SOUTHERN JAPAN'S STRATO-VOLCANO, Sakura-jima, had the country's biggest recorded eruption in 1914. Today
Sakura-jima has hundreds of moderate eruptions a year, and produces andesitic lava that forms dense, angular, smooth sided blocks. Flows of andesitic lava move very slowly down the slope, and tend to be thick and short, with a level surface unlike the broken glass texture of basalt lavas. Detached blocks form on the surface of the andesitic lava, and under the blocky top layer new, dense, andesitic lava is pushed out and forward in a big wad from the inside core, and slowly moves the flow forward.

Lava effusion rate, Mauna Loa, Hawaii, USA ▷

THE QUANTITY OF LAVA that erupts over a specific period is its effusion rate. Thin basalt lavas have higher effusion rates and travel 'fast and furious' compared to the dense, slow andesitic or rhyolitic flows. Mauna Loa's last eruption to reach Kona, now a bustling seaside resort town, was in 1955 and involved powerful lava effusion rates of 100cu m / 3,531.5cu ft per second. Today Kona has the fastest-growing population rate in Hawaii, but Mauna Loa is poised to erupt again posing a serious hazard. Thousands of holiday and residential homes, hotels, restaurants and shopping malls built on sharply inclined slopes, combined with the likely fast-flowing, high volume rivers of lava, points to a potential huge disaster.

Longest flowing channelized lava, Puu Oo Vent, Kilauea, Hawaii

HIGHLY FLUID BASALT ERUPTIONS travel great distances by forming lava channels stretching kilometers down inclined topography. Lava channels can form in both pahoehoe and aa basalt lava. As lava streams into rivers the banks cool and thicken forming a natural semi-solid levee. The hot molten center of the river keeps flowing. In aa lava channels, thick chunks adhere to the banks further insulating the central stream. The Puu Oo eruption is the longest continually flowing eruption in recorded history. It began with spraying lava fountains on an Eastern flank vent of Kilauea in 1983, lava splash back created a 300m / 985ft cone around the vent and as more and more lava was pumped out of Puu Oo numerous channels advanced downslope pouring into the Pacific Ocean at the base of Kilauea volcano.

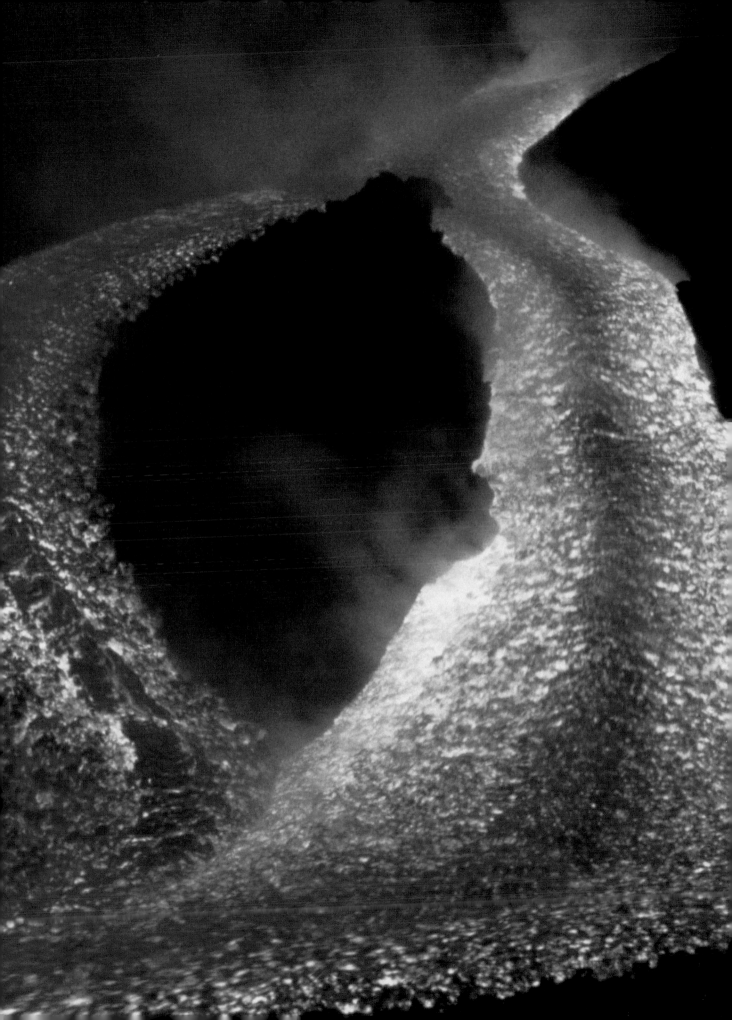

Jokulhlaup, Grimsvötn volcano and Vatnajokull ice sheet, Iceland

THE VATNAJOKULL ICE SHEET is larger than all of Europe's glaciers grouped together and covers an 8,100sq km /
3,127.4sq miles chunk of Iceland, including the active Grimsvötn volcano and its hot caldera lake, Vatnajokull. When hot
lava meets cold ice, massive floods called jokulhlaups are unleashed and in 1996 a melted jokulhlaup sloshed down the
Skeiethara river channel at 45,000cu m / 1,589,160cu ft per second, washing away homes, bridges, power lines and
roads. Several 1,000 ton icebergs cracked off the glacier's nose and plummeted 50km / 31 miles downstream, leaving
a 6km- / 3.7 mile-long ice canyon in the glacier at the jokulhlaup's origin. The tidal wave of melt-water ended days later,
scattering giant icebergs and 100 million tons of volcanic debris over the flood-washed alluvial plane.

Lava spine, Soufrière Hills, Montserrat

WHEN THICK, PASTE-LIKE LAVAS ARE SQUEEZED UP and thrust out of long, narrow openings they create dramatic, slender extrusions called lava spines. The lava spine that formed in 2002 on Soufrière Hills volcano, Montserrat, was the largest ever recorded on that volcano. The surrounding villages were put on high alert and in twilight and darkness, the glowing red spine could be seen from all over the island. Residents watched as red chunks and blocks crumbled off the lava spine near the Belham Valley and tumbled down in gobs and wads, looking like melting candle wax. When an entire spine collapses, it can cause a devastating surging pyroclastic flow.

◁ Pyroclastic flows, Vesuvius and Pompeii, Italy △

PYROCLASTIC FLOWS ARE STEALTHY, lightning-fast, sometimes silent avalanches of superheated ash, gas and rocky infernos that flash across vast areas, pulverizing everything in their way. The heat and force incinerates people, melts automobiles, levels forests and buries cities in seconds. In 79AD, after 1,000 years of peaceful slumber, Italy's Vesuvius violently burst to life. A massive pyroclastic flow buried the ancient city of Pompeii and 2,000 inhabitants in one day under 9m /30ft of grey ash, as solid as cement. In 1709 a farmer digging for a well on the site discovered the top of a building, resulting in a swarm of archeologists visiting Pompeii. These pale, life-sized, 3D body casts were made by pouring plaster into concave body pockets found under layers of volcanic ash and debris.

Carbonatite lava, Ol Doinyo Lengai, Tanzania

OL DOINYO LENGAI IN THE East African Rift System is the only known active carbonatite volcano in the world. Carbonatite lava is the rarest on Earth, and is so named because it contains more than 50 per cent carbonate minerals and less than 10 per cent silica glass. Carbonatite lavas are very cool at around 500°C / 932°F compared to basalts which can reach 1,200°C / 2,200°F. Because they lack silica, which stiffens lava flows, carbonatite lava is very fluid, thin and runny. In daylight such lavas look like flowing black-brown mud, but when seen at night they glow with an unearthly, deep rosy brilliance. The indigenous Massai people named the volcano Ol Doinyo Lengai (meaning Mountain of God) because they believe it to be a holy place where their god Engai resides.

123

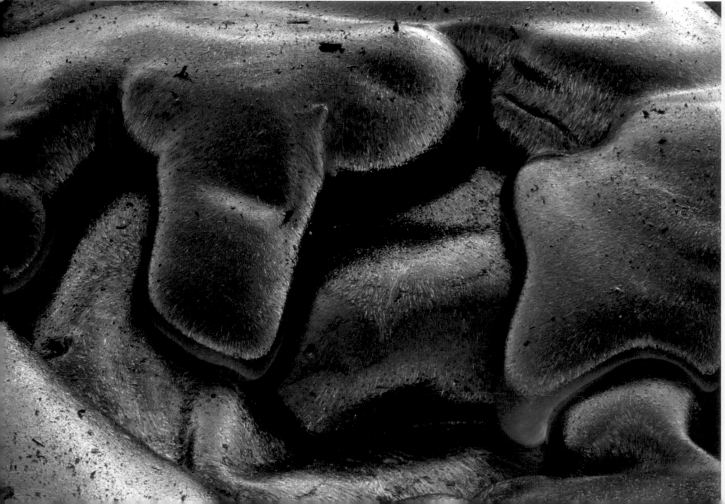

◁ Composition of lava, silver pahoehoe △

THE COLOUR AND TEXTURE OF cooling lava flows indicate the chemical composition of the magma. Lobes of basalt lava are usually 50 per cent silica glass giving readings of 900–1,200°C / 1650°F–2190°F at the moment of eruption. As the hot lava encounters the much cooler air temperatures, melted silica glass hardens and crystallizes into shiny, mineral grains that form a coarse crust, about 1cm / 0.5in thick. The shiny skin insulates the molten lava inside and these crystallized grains also reflect light giving the tops of some pahoehoe flows a brilliant silver or beautiful silvery-blue cast.

125

Lava fountains, Krafla, Iceland

ICELAND'S KRAFLA IS AN ENERGETIC volcano with spectacular, gushing lava fountains. Buoyant gas bubbles in the magma rush to the surface vent, then burst causing beautiful, explosive fountains of lava to spray into the air. Fountains average 10-100m / 33-328ft high, but some are even higher. Icelanders named the surging curtains of lava 'Krafla Fires', and have harnessed them to produce geothermal energy and provide heat and power. Krafla's lake-filled crater is 10km / 6.2 miles wide, and is aptly named the Viti Crater (viti means 'hell'). The last big eruption here happened in the 1980s. Iceland also has the world's longest, continuous lava fountain – 25km / 15.5 miles long – at the Laki Fissure in southern Iceland, which erupted in 1783.

Sierra Negra, Isabela Island, Galápagos Islands

VOLCAN SIERRA NEGRA, A SHIELD VOLCANO with an oval-shaped crater on the Galápagos National Park's Isabela Island, Ecuador, roared to life in 2005 with a loud boom heard throughout the town of Villamil. A yellowish erupting plume of gas and steam rose 20km / 12.4 miles high, and four individual vents began to issue lava at the same time. Three rushing rivers poured down the sides of the volcano and 200m- / 656ft-high incandescent lava fountains blasted into the sky to the delight of passing tourist boats. Galápagos Island volcanoes are fed by an undersea hot spot or plume of magma rising from inside the Earth.

Dome fountains, Mauna Ulu, Hawaii, USA

IN MAY 1969 THE GROUND TREMBLED under Mauna Ulu, an eastern-flank vent on Kilauea volcano, Hawaii, and lava burst out, some of it 'fountaining' 540m / 1,772ft high. 875 days of continuous outpourings followed, and eruptive activity continued for five years. The lava covered 45.3 square km / 17.5 square miles and built a dome-shaped shield mound over the vent. Mauna Ulu also produced some rare, 20m- / 65.6ft-high dome-shaped fountains not frequently seen. Lava dome fountains are hemispherical mounds of bubbling lava, occurring when the force of lava coming from the vent is equal to the mass of the ponded lava through which it erupts. These spectacular dome fountains erupted from a vent in the bottom of a ponded lava lake.

Heimaey lava fountains, Iceland

THE 1973 FOUNTAIN ERUPTION at Heimaey volcano in the Vestmannaeyjar archipelago, Iceland, illustrates the resourcefulness of the Icelanders affected by the event. A 2km- / 1.2 mile-long fissure spurted lava fountains that threatened to seal off the main fishing harbour while more than 350 buildings burned or were buried up to their rooftops in lava cinders. Some 5,000 townspeople were evacuated, but many remained and fought back using 43 water pumps to drench the creeping lava with seawater. They pumped 6,000 million liters / 1,320 million gallons onto the lava, solidifying and diverting it and ultimately the fishing harbor was saved. Residents subseqently returned, rebuilt their town and used the smoldering hot lava for geothermal energy production.

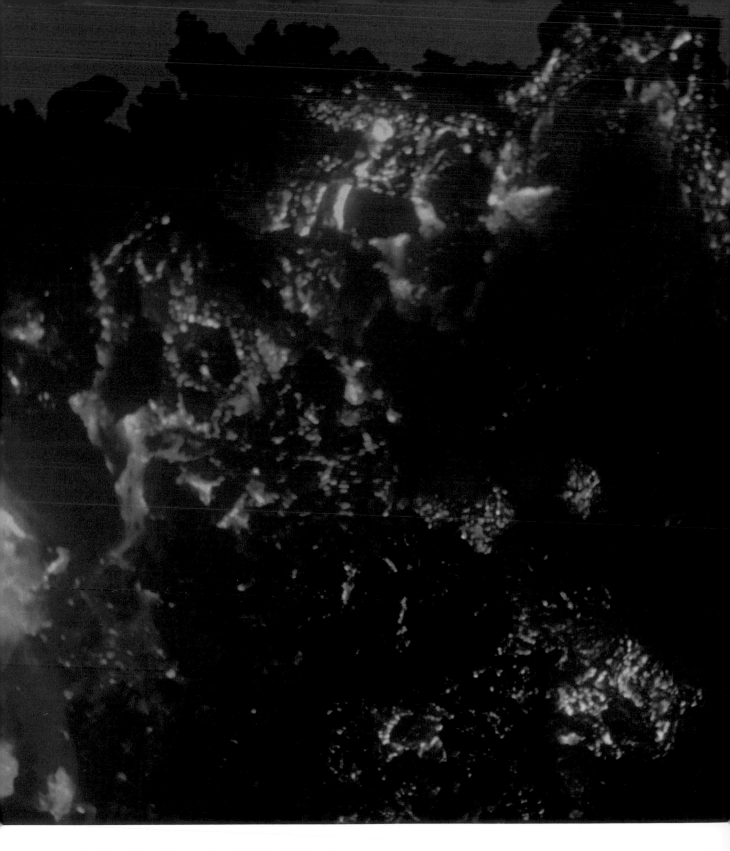

Arching lava fountains, Hekla, Iceland

ICELAND'S MOST ACTIVE VOLCANO, Hekla, is a tuya, a volcano that originally erupted under a thick sheet of ice which levelled its summit. Over time the active, flat-topped volcano melted the ice, and emerged through it. What also makes Hekla interesting is that it has unusual episodes of arching fountains which launch out from the vent, then arc sideways in the shape of a rainbow. A deformed vent or a vent on the side of a cone can cause these arching fountains. No one can predict when Hekla will erupt in the future, but it seems to follow a pattern that means the longer it remains quiet, the bigger the fountains will be when it finally does erupt.

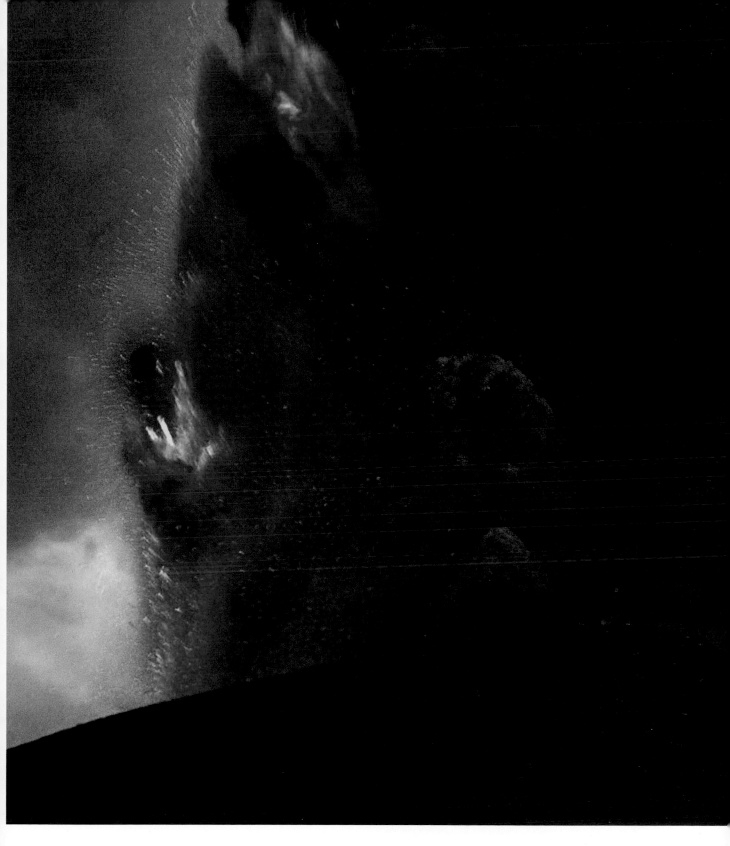

Study of volcanoes, Etna, Sicily, Italy

THE STUDY OF VOLCANOES is a new and inexact science and, though we now have sophisticated monitoring techniques and instrumentation, no one can precisely predict when, or how, a volcano will erupt. The best way for scientists to learn more is by studying erupting volcanoes. In 2000 they used a spectroscope to take infrared images of 64 sensational, jet-propelled, basalt lava fountains on Mt. Etna, and discovered that different gasses bubble and burst out of magma at different times and in different amounts, possibly determining how high the lava shoots.

Strombolian lava fountains, Stromboli, Italy

STROMBOLI VOLCANO IN ITALY has ejected beautiful fountains of lava, Mother Nature's fireworks, for centuries, so much so that Stromboli earned the name the 'Light House of the Mediterranean' because passing ships can see the fountains from out at sea. The stunning fountains are so long-lived and famous that similar fountains are called Strombolian eruptions. There are from three to ten active vents in the throat of Stromboli creating some lively, multiple eruptions. They occur every half hour or so, with bigger explosions every few years accompanied by loud blasts and showers of ash and cinders. Stromboli has been continuously erupting for over 2,000 years.

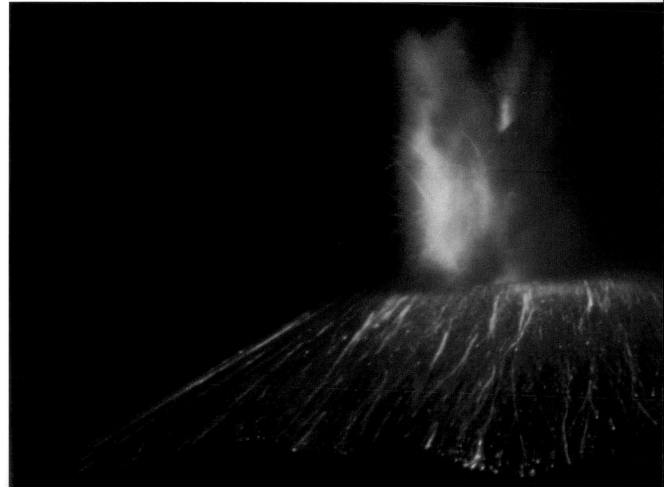

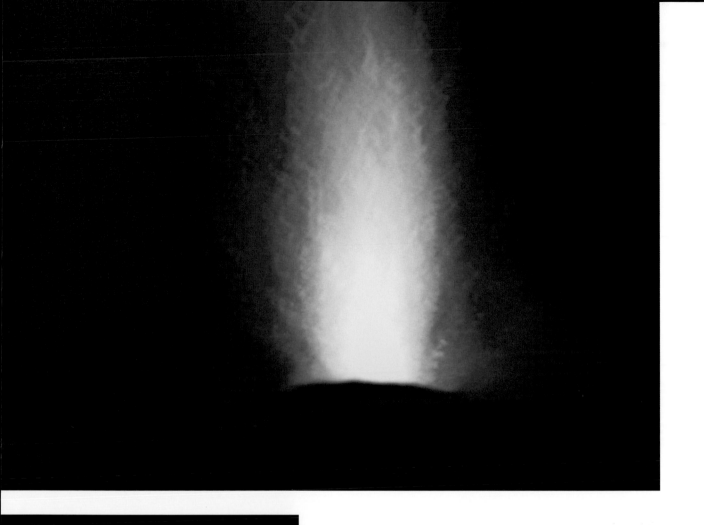

Lava fountains form cinder cones, Paricutin, Mexico

PARICUTIN IS A BASIC CINDER-CONE volcano, made from lava fountains in a Mexican cornfield in 1943, which gave watching scientists a rare opportunity to learn a lot about new volcanoes. This picture shows gas driven lava exploding in fountains from narrow vents and being blown to bits. These lava globs cool, congeal and fall to the ground as cinders in a circular mound, with a bowl-shaped depression on the top and around the vent. Paricutin's cinder cone was 400m / 1,312ft at its highest following the eruption.

133

Lava fountains, MacKenny Crater, Pacaya, Guatemala

GUATEMALA'S PACAYA IS A VOLCANIC complex consisting of two strato-volcanic cones and several dacitic lava domes. Since 1965 it has been famous for explosive fountains visible from Guatemala City, the country's capital. Lava fountains, starting small then building to powerful plumes hundreds of meters high, frequently erupt from MacKenny Crater, named after a local man who has been making annual climbs to Pacaya's summit for decades. In the 1980s and early 1990s tourists climbing Pacaya were frequently assaulted by bandits who roamed the slopes, robbing climbers of money and cameras. Now Pacaya is a relatively safe park where people go hiking and climbing.

Youngest volcano in Costa Rica, Arenal

THE SYMMETRICAL ANDESITIC CONE OF ARENAL, in Costa Rica, has major, explosive eruptions every few hundred years. After a 400-year sleep, Arenal blasted to life in 1968 with a lethal pyroclastic eruption tossing large blocks of lava 5km / 3 miles high. Since then periodic lava fountains and double fountains have delighted tourists at the towns of Tabacon and Fortuna. Arenal also produces lava balls up to 25m / 82ft in diameter that drop out of the fountains and roll downhill. The lower slopes of Arenal are blanketed in pristine rainforest where sloths, wild parrots, howler monkeys, hundreds of colourful bird species and the giant morpho blue butterfly live, and the indigenous Guatuso Indians have also lived beside Arenal for centuries.

Lava fountain induced snow avalanche, Villarrica, Chile

HUMAN LIFE SPANS ARE MINISCULE when compared to the millions of years a volcano can be active. When volcanoes are resting or dormant they can be mesmerizing, and people forget that they have the potential to be deadly. Villarrica is a perilous, snow-covered strato-volcano in the town of Pucon, Chile, where there is a booming ski centre. Guides take groups up to the summit crater, undeterred by the lava fountains that frequently melt channels in the summit snow, causing avalanches of ice, snow and lava rubble. As recently as 1964 there was a crater-lake eruption that melted the ice creating mud lahar flows killing several dozen people.

Volcanic lightning, the great Talbachik fissure eruption, Russia

THE KAMCHATKA PENINSULA IN RUSSIA is sprinkled with about 40 volcanoes. One of these, Talbachik, is famous for its powerful jets of incandescent lava fountains that erupted in 1975 from a new fissure along its south rift zone which became known as 'the great Talbachik fissure eruption'. Violent volcanic updrafts in blasts like this, that eject lava jets and ash plume particles, are saturated with an electrostatic charge caused by the friction of millions of volcanic grains glancing off each another in the blast. This sometimes causes sensational lightning spectacles within the eruption plume such as those displayed by Talbachik. The great Talbachik fissure eruption was the largest basaltic lava eruption in the 10,000-year history of all the Kamchatka volcanoes.

Highest lava fountains on earth, Kilauea Iki, Hawaii, USA

KILAUEA IKI (MEANING 'LITTLE KILAUEA') is a pit crater formed by a collapse near the summit of Kilauea volcano and the 1959 lava fountains at Kilauea Iki, at over 600m / 1,968ft, are the highest ever witnessed (the fluid basalt lavas of Hawaiian volcanoes create magnificent, pulsing lava fountains). They resulted in a big cinder-cone named Puu Puai (or 'Gushing Hill') forming at their base. Lava from the fountains formed in pools, and then was then locked inside the pit crater when the crust – which is now about 30m / 98ft thick – cooled. Scientists have drilled a deep core sample of the hardened crust to study the minerals forming as it cools, and they monitor the hot lava that remains deep in the pond using seismic equipement. Meanwhile hikers continue to enjoy walking over the crust amidst steaming vents.

◁ Nyiragongo lava lake, Zaire △

LAVA LAKES FORM IN CRATERS and depressions, and can be completely molten or partially solidified. Nyiragongo, part of the Virunga volcanic chain, is a steep-sloped strato-volcano home to a spectacular lava lake that has sloshed in its deep summit crater for the past 50 years. In 1977 the lake drained unexpectedly in less than one hour. Flows poured out the sides of the volcano at 60km / 37 miles per hour killing 70 people, then in 2002, Goma, a haven for millions of refugees from the Rwandan civil war, was hit by lava from nearby Nyiragongo. Scientists rushed to the site to search for clues to predict future volcanic eruptions. The lake has since refilled and the top crust continually cools, turns black, and is then ripped open in red seams by lava updrafts.

139

Kelimutu Volcano Lake, Flores Island, Indonesia ▷

LONG AGO DIFFERENT MINERALS DISSOLVED in three summit crater lakes at Kelimutu volcano, giving them brilliant colours that change hue due to the fluctuations in geochemical content. Tiwi Ata Mbupu ('The Lake of Old People'), on the western edge, is azure. Tiwu Nua Muri Kooh Tai ('The Lake of Young Men and Women'), which is the deepest lake and shares a volcanic wall with its sister lake, is emerald and occasionally streaked with bright yellow sulphur foam. Tiwu Ata Polo ('The Enchanted Lake'), on the south-east ridge, is brick red and orange. Underwater fumaroles cause each of the lakes to upwell, bubble and occasionally cause phreatic eruptions leaving a rim of white froth at their edges.

Sulfur mud volcanic lake, Vulcano, Italy

EARLY ROMANS BELIEVED Vulcano volcano in Italy's Aeolian Islands to be the entrance to hell. Today it is a popular summer tourist resort, and people come from all over the world to this tiny island volcano, with one of the prettiest black lava sand beaches in Italy, to bathe in its smelly, mustard-coloured, sulphur mud lake that's said to help cure aches, pains and various skin afflictions. Locals have named the rotten egg-reeking mud baths Laghetti de Fanghi. Hot volcanic gasses also bubble up nearby in the ocean where bathers wash off the gunk, but the smell remains on the skin for days. An impressive steaming crater – the Gran Cratere – also sits atop Vulcano, which has had seven recorded major eruptions.

Pinatubo crater lake, Philippines

AFTER 635 YEARS OF REPOSE Pinatubo, on the island of Luzon, blasted to life in 1991 with eruptions of lahars, pyroclastic flows and ash. So much sulphuric acid and volcanic aerosols were blasted into the atmosphere that the global temperature lowered by 0.5°C / 0.9°F. Monsoon rains following the eruption quickly refilled the jade summit crater lake, which was acidic and had a temperature of about 40°C / 104°F (it has since cooled and diluted). In 2001 it threatened to overflow and flood, so workers cut a notch in the crater wall and successfully drained the lake to a safer level. Pinatubo's jungle slopes have been populated for centuries by the indigenous hunter-gatherer Aeta people, but their traditional life was severely threatened by the disruption and damage from the 1991 eruption.

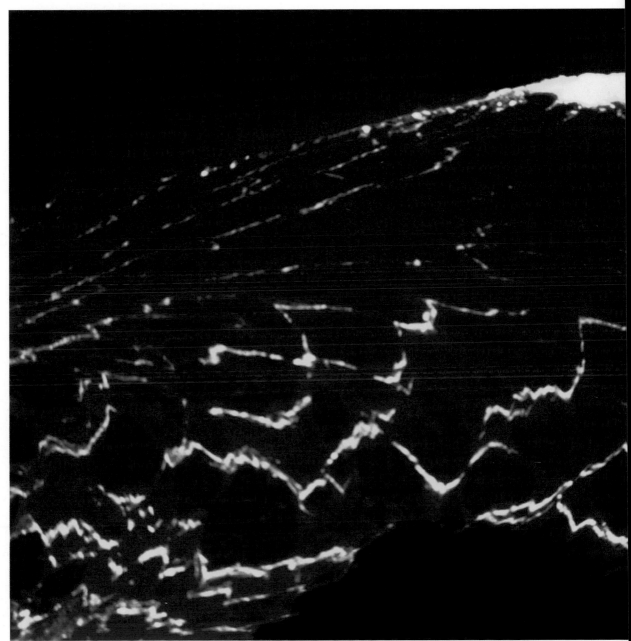

Kupaianaha lava lake, Kilauea, Hawaii, USA

CALDERAS AND CRATERS ON VOLCANOES are sunken depressions into which lava flows, forming molten lava lakes, some of which form over sunken vents, as at Kupaianaha in Hawaii. Lava lakes have an intrinsic beauty, but if a lake wall breaks the lava can escape. Voluminous outpourings of lava fed the Kupaianaha lake until it over-flowed and poured away down slope for five years from 1986 to 1992, into the Pacific Ocean. Meanwhile, if gasses escape from the lake floor vents they cause explosions, upwelling and lava waves. As the air cools it, the top skin of the lava solidifies, ending up with jagged edges that look like scales of fish, floating on the lake surface, or reminiscent of the Earth's tectonic plates moving above the mantle.

Lake Nyos disaster, Republic of Cameroon

LAKE NYOS, FORMED BY AN ERUPTION 500 years ago, sits in the Mt. Oku volcanic field of western Africa. All magma contains gas bubbles and, at Nyos, large pockets of carbon dioxide (CO_2) lie dissolved in the lake water. When the lake gets rough and churns over, the gas bubbles burst and CO_2 is shot into the air, as happened in 1986, when a catastrophic release of magmatic CO_2 from the volcanic lake killed 1,700 people. Scientists are still studying the gas release mechanism at Lake Nyos and, in 2001, officials used a large pump to try and release some CO_2 to avoid a future catastrophe. Lake Monoun, southeast of Nyos, also contains volumes of CO_2.

Cold water lake, Lake Botos, Poas, Costa Rica

WHEN UNDERGROUND MAGMA RESERVOIRS suddenly drain, bowl-shaped depressions, called craters or calderas, are filled by rainwater and/or overflowing rivers. Botos Laguna is a cold, cobalt blue rainwater lake on top of Poas volcano in Costa Rica and is surrounded by lush rainforest vegetation, in stark contrast to the harsh, desolate region surrounding its twin lake, the hot Laguna Caliente. The rainforest contains 79 species of colourful birds including hummingbirds, toucans, cuckoos, tanagers and the sensational, long-feathered, golden-green quetzal bird which help to draw many visitors who hike along the Escalonia Trail, through the forest, past cypress and oak trees, hanging mosses and orchids, to snatch a look at this beautiful lake.

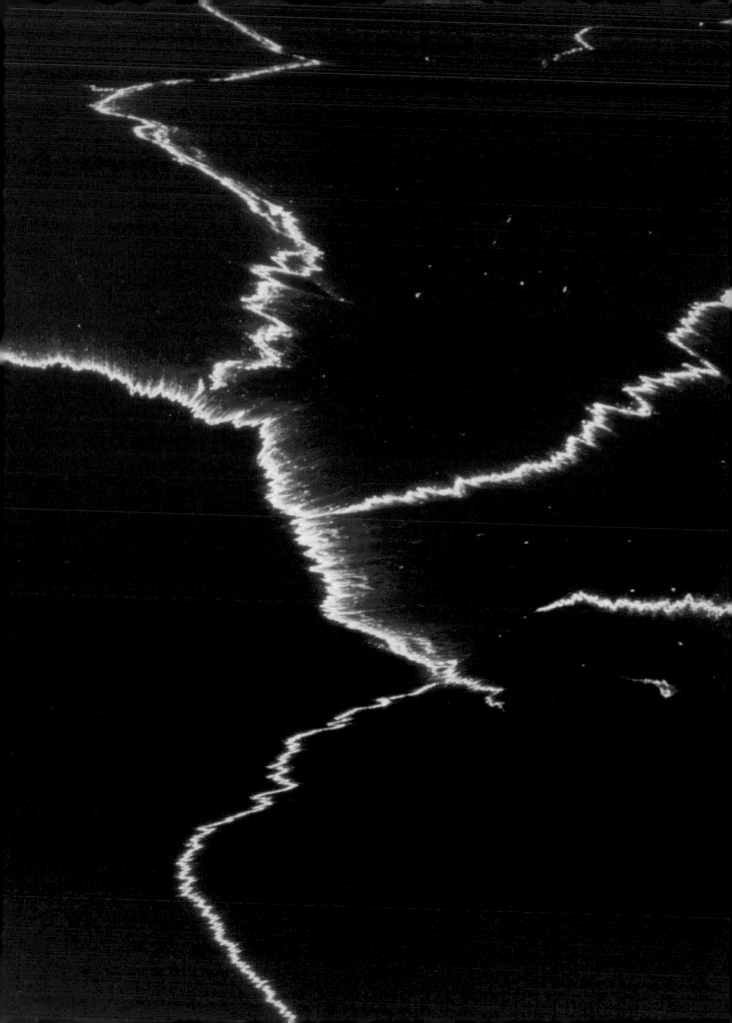

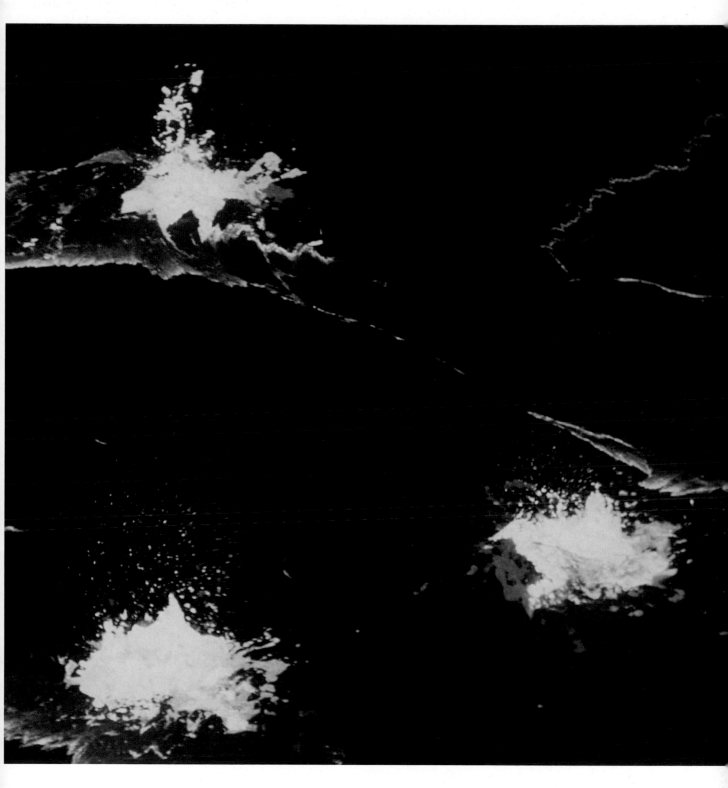

◁ Oldest lava lake, Erta Ale, Ethiopia △

ERTA ALE, A SHIELD VOLCANO in East Africa, has a vigorous lava lake (the lava composed of basalt and rhyolite) in its summit crater that may well be 90 years old. The cooling of the crust of the lava lake blackens the lake and as heat builds up beneath the crust covered lava, convection tides tear and rip the surface apart once more. In 2003 the lake blasted up lava fountains and pumped large amounts of sulphur dioxide (SO_2) into the air. The Afar Triangle of Africa, where Erta Ale is located, is hot, remote and wild, making close monitoring of the volcano very difficult.

Lava lake and lava falls, Alae Lai, Mauna Ulu, Hawaii, USA

MAUNA ULU, A FLANK VENT on Kilauea volcano in Hawaii Volcanoes National Park, began erupting in 1969 and had 875 days of continuous outpourings of lava. Two pit craters filled to the brim with sloshing lava and the tops of the lava rivers cooled and turned to lava tubes. One of these tubes continually fed a lava lake, Alae Lai, creating a rushing lava fall. The National Park opened the lake to lucky visitors who stood at the edge of the burning caldron and peered into it. In 1974 the Mauna Ulu eruption ended and the lake drained for good.

Acid lake, El Chicon, Mexico

EL CHICHON IS THE LAVA DOME of a complex volcano in Chiapas, Mexico. In 1982 a powerful sulphur-laden eruption blew apart the dome, pyroclastic flows killed hundreds of people and 100km / 62 miles of pristine rainforest was decimated. A 1km- / 0.6 mile-wide crater left by the blast was filled by ground water, which became an acid lake. The eruption also fired a large volume of sulphur dioxide dust into the Earth's stratosphere which circled the globe for three years causing exceptionally colourful sunrises and sunsets as far away as Alaska. Before 1982 El Chichon was believed to be extinct proving that volcanoes are always unpredictable.

△ Quilotoa Caldera Lake, Ecuador and Telaga Warna Lake, Indonesia ▷

COLOURFUL VOLCANIC CRATER LAKES form when water is trapped in low craters and calderas that are not erupting lava. The water can be fresh and full of acidic chemicals injected from fumaroles below, but it can also come from monsoons, rain, underground water supplies, rivers and streams. The turquoise Quilotoa Caldera Lake in Ecuador's Andes Mountain Range is 3km / 1.8 miles wide, 250m / 820ft deep, and 2,270m / 7,447.5ft above sea level. It is fed by a hot spring along the eastern ridge. Telaga Warna Crater Lake in Java bubbles sulphur gasses to the surface giving the entire lake area a rotten egg smell.

Most acidic lake, Laguna Caliente, Poas, Costa Rica

UNLIKE ITS COLD-WATER SISTER-LAKE, Botos Laguna, Laguna Caliente is hot – up to 960°C / 1760°F. It is assaulted by underwater fumaroles blasting out boiling gasses (SO_2, CO_2 and HCl), making it the most extreme, acidic and sulphuric lake in Central America. These harsh gasses, when carried by the winds, have de-forested the north-western slope of Poas volcano in Costa Rica. Scientists converge on Laguna Caliente sampling the lake's unique chemicals in an attempt to unravel hydrothermal secrets. For decades they had only been able to theorize whether molten sulphur was at the bottom of active volcanic lakes but when Laguna Caliente dried out in the 1980s, their theory was finally proven to be correct.

Crater Lake, Mt. Mazama, Oregon

WHEN A VOLCANO ERUPTS huge amounts of magma, the remaining void can collapse in on itself creating a large caldera. Seven thousand years ago an explosive eruption suddenly emptied a giant magma storage chamber under Mt. Mazama in Oregon, USA, deflating the summit and leaving a crater 8km / 4.9 miles x 10km / 6.2 miles wide. Over time snow melt and heavy rains filled the depression, and now the 600m / 1,968.5ft deep, blue Crater Lake is the second deepest lake in the USA. It has an ancient cinder cone called Wizard Island peeking out of the surface. Collapse calderas can form in as little as a few hours.

Taal Volcano Lake, Philippines

THE HUGE LAKE TAAL is 15km / 9.3 miles wide and 20km / 12.4 miles at its length, and occupies the interior of Taal caldera just 3m / 9.8ft above sea level. The Philippines' Taal volcano is powerfully dynamic and its lake holds a volcanic island that has been the source of eruptions and deadly pyroclastic flows which can sweep horizontally across the top of the lake and smack into populated regions, killing many people. Some of Taal's blasts even create tsunamis within the lake. Taal has recently started showing show signs of unrest once again.

Boiling Lake, Morne Watt, Dominica, West Indies

EAST OF ROSEAU, DOMINICA'S CAPITAL CITY, Morne Watt volcano's Boiling Lake thermal region, near the ash-laden Valley of Desolation in the Morne Trois National Park, is a popular tourist attraction for extreme adventurers. Getting to the lake involves a hair-raising day-long climb along a rugged, narrow, steep trail. On arrival you are greeted with the site of Boiling Lake itself – a vent of bubbling sulphur water enveloped in a vaporous milky liquid cloud that is actually a flooded fumarole about 90m / 295ft wide. Morne Watt last erupted about 1,300 years ago.

155

Maly Semiachak crater lake, Kamchatka, Russia ▷

MALY SEMIACHAK'S SCALDING CALDERA lake is a startling robin's-egg blue, and scientists have found clues to its radiant azure colour in the dissolved chemicals in the lake itself and the fumaroles that line the lake's side, which reach temperatures of 900°C / 1,650°F. The Russian lake sits inside Troitsky caldera, which formed about 400 years ago and the compound strato-volcano they sit atop last erupted in 1952 and is still active today. Most of its eruptions have been explosive Vulcanian blasts, with the most devastating of recent history in 1804. It's so popular that a Russian postage stamp with its image was issued in 2002.

Waimangu Volcanic Valley Lake, New Zealand

MOUNT TARAWERA ERUPTED in New Zealand in 1886, demolishing a large part of the North Island. A 17km / 10.5 mile-long volcanic rift cracked the mountain in half forming a string of fuming craters, two of the most amazing hot springs in the world – the colourful Frying Pan Crater Lake and Inferno Crater lake – as well as geysers, hot-water swamps and lively geothermal activity. The region was named Waimangu Volcanic Valley, and its unique and impressive hydrothermal activity is still fuelled by Tarawera volcano. Waimangu has rare thermal plants which have adapted to its alkaline and acidic soil, and the whole area is intensely studied by scientists who have learned that the Earth's crust is only 10km / 6.2 miles here, compared with the 25km / 15.5 miles in non-volcanic areas.

Sulfur mining, Kawah Ijen volcanic lake, Java, Indonesia

MANY VOLCANIC LAKES CONTAIN deep pockets and deposits of smelly, yellow, heavy sulphur that sink to the bottom in thick layers, having erupted from underwater fumaroles. The Ijen volcano complex in Java is home to the world's largest, extremely acidic crater lake, Kawah Ijen. It is 1 km / 0.6 miles wide, and streaks of bright yellow sulphur dust criss-cross the bright turquoise surface. Each day heavy double baskets of sulphur are draped over worker's shoulders at the local mine, and are hauled by hand up from the lake crater floor. Once processed, the sulphur is sold for industrial fertilizer and dyes.

Lava pond inside vent, Puu Oo, Kilauea, Hawaii, USA

THE PUU OO CONE IN HAWAII contains an active lava pond. Like a bath, it fills, drains and refills with lava again and again but, occasionally, it overflows and drips down the sides. The upsurges in Puu Oo lava pond are preceded by shallow earthquakes that occur when magma moves into the pond, shredding the dark crust. In the early 1990s the beautiful coastal town of Kalapana was buried under molten lava from a Puu Oo eruption. Tourists come to see the orange-pink glow on the underside of clouds above the vent.

Aa lava flows

AA LAVA IS CHARACTERISED BY A JUMBLED, ragged appearance compared to the smooth-flowing pahoehoe. Although slightly thicker than pahoehoe, aa can flow faster than pahoehoe on a steep incline. The front of the flow moves along like a tractor tread rolling continually forward, and the top is covered by rubble and jagged cinder chunks which insulate and keep the lower portion fluid. As the flow moves on, big, ragged clumps of the cragged top break off and fall forwards. Sometimes a pahoehoe lava flow converts to aa due to cooling, an increase in viscosity or thickness, de-gassing or crystallization, but an aa lava flow never becomes a pahoehoe lava flow.

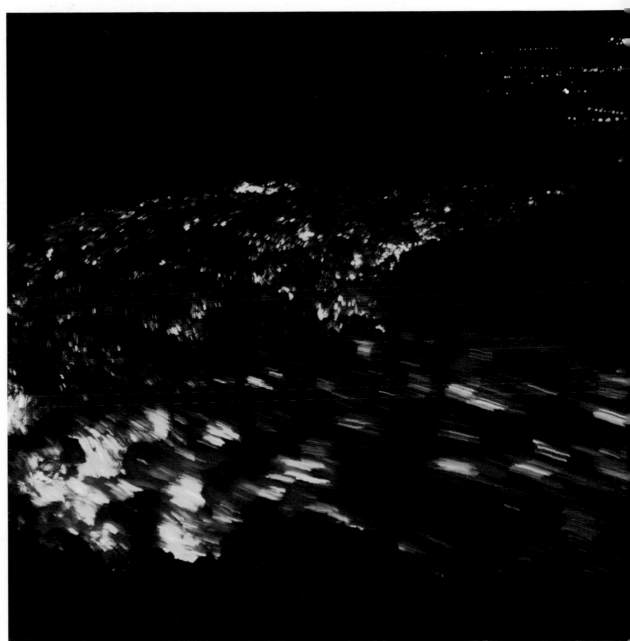

FLOWS, FOUNTAINS & LAKES

△ Pahoehoe lava flows ▷

THE TOP OF A THIN PAHOEHOE LAVA flow is pliable like a plastic wrap. As it cools it becomes cumbrous and is pulled, squeezed, rippled and tugged by the faster moving, hotter lava underneath. This stretching creates some extraordinary intricate surface shapes. Depending on the temperature, composition and speed, the supple surface can be ropey, wrinkled, crumpled, ribbed and rippled, forming beautiful patterns when hardened. Usually the front of the flow will advance in a bulbous lobe or toe shape with crinkles on top. As the lava encounters vegetation or trees, flames erupt as the obstacle is burned and smothered. Sometimes the tops form a silver crust and then crack open, and new red lava oozes out like toothpaste.

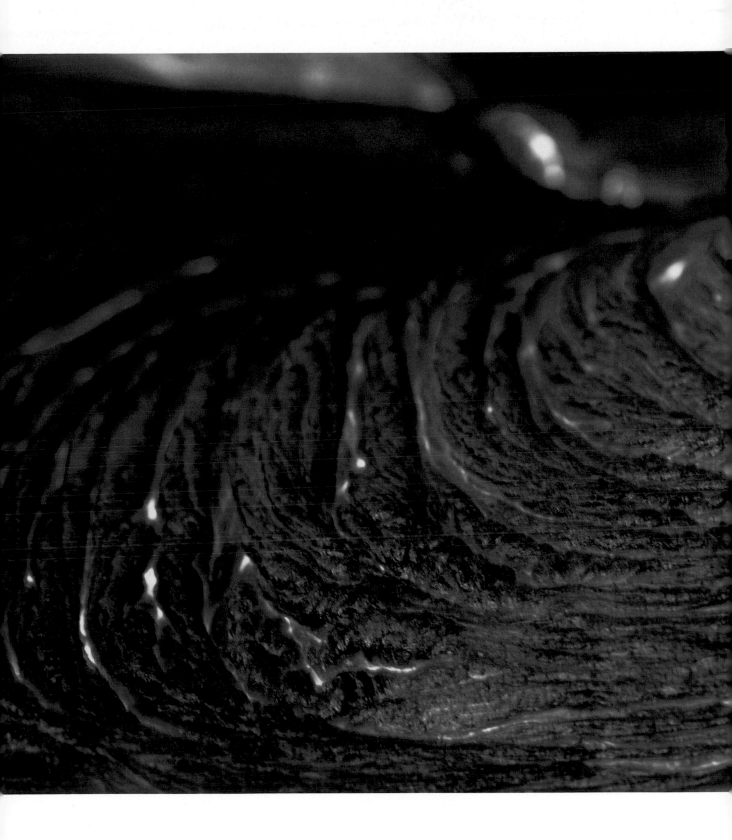

EVERYONE IN THE WORLD WOULD RECOGNISE A CLASSIC POINTED CONE spewing lava from its top as a volcanic land form. However, few people could immediately identify a circular coral atoll filled with translucent turquoise water, as the submerged, eroded cone of a once-active volcano. Neither would they recognise a green-sand beach as evidence of volcanic olivine rock, derived from molten lava, cooled and hardened by the sea, then eroded by the waves and deposited on a beach. As well as the familiar dramatic landscapes that have been shaped, destroyed or created through past eruptive activity, this chapter examines the more unusual landscapes and geological features that have arisen due to volcanism. Planet Earth is cloaked with evidence of tectonic plate movement. From the eroded and now tranquil Kohala volcano in Hawaii whose flanks, now carved with gorgeous chasms, are rich in wildlife and belie their fiery origin, to vast spreading rift valleys that rupture under the sea, such as in the Gulf of Aden. Scientists monitor the plentiful evidence across the Earth of the movement of plates and deformation is the name given to the actual movement of points of the Earth, a result of the build up of pressure in the the Earth's brittle crust. This study of the build up and transfer of 'stress' from one area of the crust to another could yet prove vital in understanding how and where earthquakes and volcanoes erupt. Volcanic land forms are the result of plate tectonics but as well as being dramatic and vast, the evidence of volcanism can be pretty or spell-

binding. Active geological processes cause intricate formations, such as at Cotton Castle in Pamukkale, Turkey where geothermal deposits build fairy tale terraces. Picturesque mineral ponds in New Zealand are coloured by deposits of gold, nickel,and copper dissolved and released through volcanic products. Even precious stones like diamonds owe their existence to internal volcanic processes. Lava flows spare beautiful green islands of trees, called 'kipukas', leaving them in their wake, dazzling amid the torched landscape. Underground tubes that once contained hot flowing lava rivers eventually dry out and become lava caves. Even extinct extinct volcanoes also give rise to beautiful landscapes and varied land forms. Wind, rain and ocean waves erode and strip away their remaining cones, leaving formations such as Strombolicchio in Italy, which is an exposed, hardened magma plug jutting out of the sea.

VOLCANIC LAND FORMS

4

Lava stalactites

WHEN PAHOEHOE LAVA FLOWS move through underground tubes, the cooled, hardened ceiling of the tube acts as insulation preventing any heat from escaping. When the roof of a lava tube remelts from the build up of heat gravity tugs the gooey lava, and elongated, pointed drips form hardened spears of rock. Unlike mineral cave stalactites, lava stalactites do not continue to lengthen when formed. When the tube drains it becomes an empty lava cave with a ceiling studded with fragile lava stalactites.

Previous pages | Collapse scar, Mt. St. Helens, Washington, USA

THE DISPLACEMENT OF VAST AMOUNTS OF EARTH and volcanic debris in the 1980 eruption of Mt. St. Helens resulted in the huge amphitheatre-shaped scar that we see today on the North side of the volcano. Prior to this eruption, Mt. St. Helens had a cone-shaped summit.

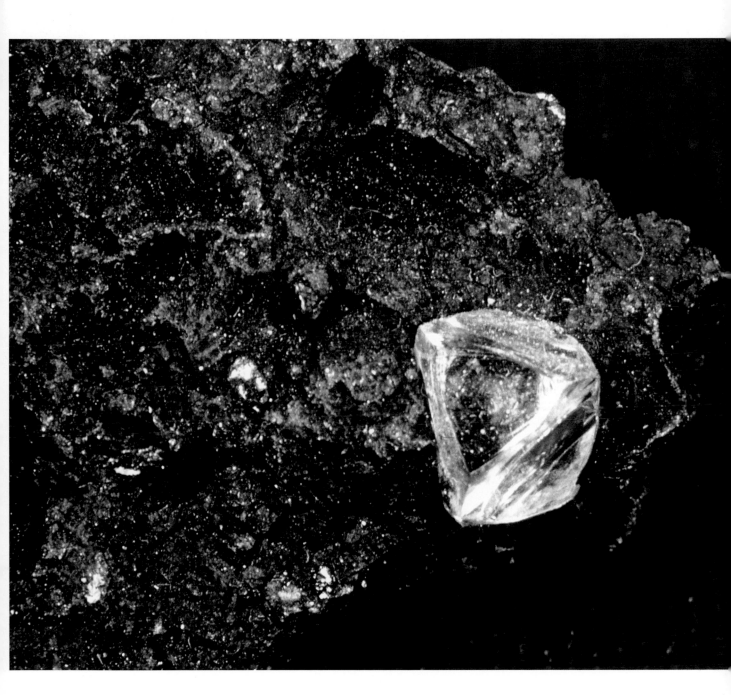

Diamond Kimberlite pipes

PRECIOUS DIAMONDS ARE GAS-BLASTED to the surface from deep within the Earth, being flung up through Kimberlite pipes. Diamonds are a form of tightly bound, pure crystalline carbon created more than 150km / 93.2 miles below the surface, where extreme pressure and heat exert the force needed to create them. To reach the surface and retain their hard, translucent composition, the diamonds must cool at breakneck speed. Scientists believe that deep, funnel-shaped Kimberlite pipes are blown violently open when gas, probably methane, from Earth's upper mantle, explodes and blasts so forcefully that the fully crystallized diamonds are thrust to the surface intact. Kimberlite pipes are named after the town of Kimberly in South Africa where the first one was found.

167

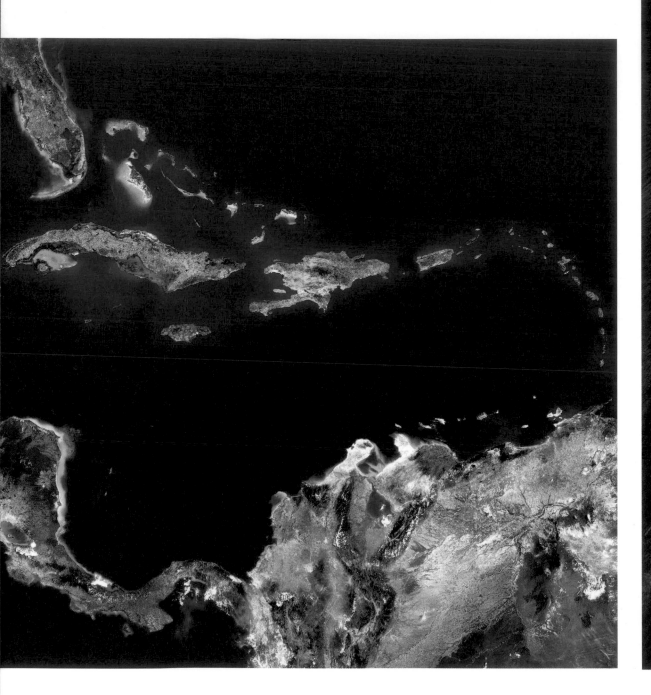

Oceanic island arcs

OCEANIC VOLCANIC ARCS ARE SPAWNED by tectonic plate subduction, when one plate slumps under another. When this happens, a plate melts into magma that erupts along the fractured plate seam creating an oceanic ridge, which ultimately rises past sea level. The Caribbean island arc contains over 7,000 islands and Indonesia and South-East Asia contain some of the world's largest island archipelagos, with thousands of islands stretching 5,000km / 3,107 miles, forming the western slice of the Pacific Ring of Fire. Many of these islands are dotted with active, smouldering volcanoes. Their isolated location, south of Asia and north of Australia, is the reason for their unique species of flora and fauna, including Komodo dragons, giant insects and strange orchids, found nowhere else.

Plate boundary volcanic island, Vanuatu, South Pacific

MOST VOLCANOES POP UP ABOVE the boundaries of the Earth's tectonic plates, a quilt of about a dozen moving slabs that bump into, spread apart or rub against each other at about the same rate that fingernails grow. The South Pacific islands, once called the New Hebrides and now Vanuatu, are a string of volcanic islands at a plate boundary. Under this chain the Pacific plate moves under the Australian plate, the latter moving at about 9cm / 3.5in per year in a north-east direction, melting and sending magma to the surface. Nine of Vanuatu's main islands are active volcanoes, and Yasur has been erupting since even before Captain Cook spotted it spewing ash in 1774. No one knows exactly how long these islands have been settled, but pottery dating back 4,000 years has been dug up on a nearby island.

Pulvermaar lake, Germany

SCIENTISTS FLOCK TO GERMANY'S VOLKANEIFEL, also called the West Eifel Volcanic Field, to study its famous maar volcanism. Volcanic maars are shaped when hot magma infiltrates ground water at shallow depths creating steam explosions. What's left after the blast is a flat-floored, low-rimmed, bowl-shaped crater. Unlike tuff rings, maars don't contain any newly erupted volcanic materials. Over 75 maars have been discovered in Germany (maar is German for lake), and Pulvermaar, a water-filled maar near the town of Gillenfeld, holds the deepest crater lake at 72m / 236.2ft. Geologists believe that the Vulkaneifel may well have more volcanic eruptions. The area is part of a European network called Geoparks, promoting geo-tourism.

Phantom Ship Rock, Crater Lake, Oregon, USA

PHANTOM SHIP ROCK, poking its gaunt 'mast' 51m / 167ft above the deepest lake in North America – the 597m- / 1,958.6ft-deep Crater Lake in Oregon – is the nose of a silica-rich, andesitic, lava flow ridge. Over 1,354cm / 533in of snow per year fill the lake in the ancient crater, which is all that remains of the ancient Mt. Mazama volcano. Lava that shaped Phantom Ship Rock was probably erupted when the volcano blew its top and then collapsed in on itself, and may even pre-date the eruption that caused the vast crater. Some of the oldest lava flows in the lake are about 400,000-years old. President Roosevelt made Crater Lake and its Phantom Ship Rock a national park in 1902.

Mt. Kilimanjaro, Tanzania

TANZANIA'S MT. KILIMANJARO (which inspired Ernest Hemingway to write *The Snows of Kilimanjaro* in 1936) is a strato-volcano and at 5,895m / 19,340.5ft high it is the tallest mountain in Africa. The summit has three peaks named Shira, Kibo and Mawenzi and portions of the summit are draped with the Furtwangler glacier. Thousands of climbers make the difficult climb to the icy summit each year. even though it's a dangerous place to be (despite not being currently active). The fumarole-stippled summit had a rockslide that killed four people in 2006. Indigenous people farming the fertile soils at the base of the great mountain have many legends and myths about demons and gods inhabiting the volcano. Scientific study has revealed a magma storage area under the volcano at a depth of about 400m / 1,312ft.

Lake Naivasha, East African Rift Valley

AS THE EARTH'S TECTONIC PLATES SLIDE against each other, weakened areas splinter and crack. Two long cracks formed on the sides of the African Rift Valley and the land in the middle, sunk and flattened. The Great Rift Valley is 6,000km / 3,728miles long, and Lake Naivasha is Kenya's biggest fresh water lake. The land beside Lake Naivasha, in the East African Rift Valley, was once a landing strip for passengers visiting Nairobi prior to 1950 but in recent years the rich, volcanic soil and temperate climate have given rise to a thriving vineyard industry. Tourists swim in the cool waters, while bird watchers and wildlife experts come to this biodiversity hot spot to catch glimpses of over 400 species of birds, including pink flamingos, as well as hippos, giraffes and monkeys.

Fumarole field, Colima, Mexico

FUMAROLE FIELDS ARE AREAS OF SIZZLING VENTS spitting vapour and gasses, that form in regions of intense geothermal activity where trapped gasses leak from shallow deposits of hot magma or igneous rocks and penetrate the surface or contact ground water. Fumaroles saturated with sulphur are called solfataras, while fumaroles rich in carbon dioxide are called mofettes, and geologists analyze the chemical makeup of fumarole vapours to learn about past and future eruptions. Colima, Mexico's most active volcano, home to a vast fumerole field, has been rumbling for millions of years, and massive debris avalanches have occurred here in the past. Today thick magma has been piling up in a dangerous dome that could collapse into a lethal pyroclastic flow, while more than 300,000 people live nearby.

Devil's Tower, Wyoming, USA

STUNNING DEVIL'S TOWER, a sacred site for native Americans, rises 300m / 984.2ft over the Wyoming plains. Sixty million years ago there was an underground intrusion, or surge, of magma into the surrounding rock, that cooled slowly into the giant, ridged block. Over time, glacial erosion of the surrounding soil exposed the natural structure jutting above the ground, its flattened summit being about the size of an American football field. Native Americans named it Mateo Tep, which means Bear Lodge. The tower is popular with climbers, but at various times they're prohibited so that native ceremonies can be held. Devil's Tower was America's first National Monument.

World's longest mountain range, The Andes, South America

THE WORLD'S LONGEST MOUNTAIN RANGE, the Andes, stretches across Columbia, Ecuador, Chile, Bolivia, Argentina, Peru and Venezuela for over 8,000km / 4,970 miles. The Andes are the aftermath of plate tectonics. As the Nazca Plate has been subducted under the South American Plate, the terrain has been lifted, and been pleated and folded. The continuous friction of the plates has also resulted in numerous active volcanoes in the range, with the snow-covered Cotapaxi, in Ecuador, being one of the highest active volcanoes. In the 1400s the Inca civilization thrived in the Andes where they constructed roads, aqueducts and the famous site of Machu Picchu.

Volcanic remnant, Strombolicchio, Italy

THE VOLCANIC REMNANT STROMBOLICCHIO is a rocky outcrop accessible only by boat, protruding from the blue Thyrennian Sea just offshore from the Italian volcano, Stromboli. An ancient satellite volcano rising from the sea floor sent a plug of lava up through its central conduit that solidified into a jutting, volcanic neck. Over time wind and waves eroded this much larger volcanic edifice into the startling remains of Strombolicchio that we see today. The volcanoes in the nearby Aeolian Islands are a classic example of the result of a fracturing of the Earth's crust, allowing fluid basalt magma to escape. The Italian navy maintains a solar-powered lighthouse on top of Strombolicchio.

Strato-volcano, Kronotsky, Russia

STRATO-VOLCANOES INCLUDE SOME OF THE MOST BEAUTIFUL and imposing volcanic structures on Earth. In fact just over half the Earth's volcanoes are strato-volcanoes erupting viscous andesitic and dacitic lava. These gas-rich, thick lavas plug the volcanoes plumbing systems for long periods, creating tremendous pressure before they blast sky-high. The pointed cones of strato-volcanoes are made of layers, like a cake, with alternating lava, ash and pyroclastic materials. If the summit is ice- and snow-covered, as at Kronotsky in Russia, the eruptions often cause mudflows (called lahars) and massive collapses and avalanches. Like Mt. St. Helens, Kronotsky is also noted for its sideways, explosive eruptions. Beautiful, pointy Kronotsky overlooks the blue of Lake Kronotskoe and has been designated a World Heritage Biosphere Reserve.

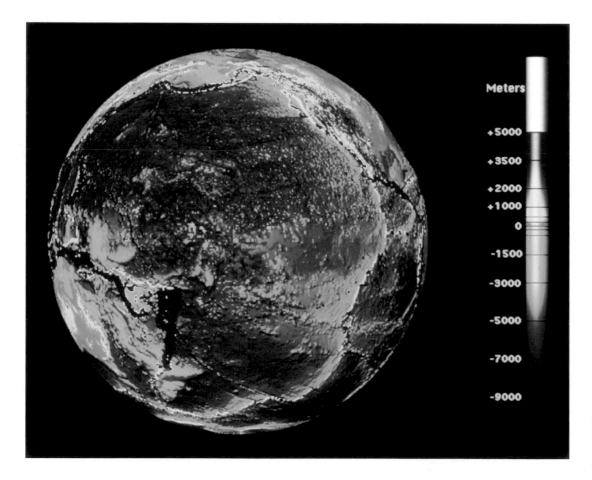

Meters
+5000
+3500
+2000
+1000
0
-1500
-3000
-5000
-7000
-9000

Pacific Ring of Fire

THE EARTH'S STIFF EXTERIOR IS COMPOSED of about a dozen plates or slabs about 80km / 49.7 miles thick. As the brittle plates shift, they snag, pull away, grind and crash into each other. Chains of volcanoes in the Pacific Ring of Fire are produced by subduction, when one plate moves under another. Three-quarters of the world's volcanoes pop up somewhere on the 'ring', and it spans the globe from New Zealand to Alaska. On this computer-enhanced image, the red dots signify earthquake epicenters, and the yellow lines indicate the huge ring of volcanic and seismic activity surrounding the large, central Pacific Plate, in the blue center. Hundreds of plate-boundary volcanoes have risen at these volatile, convergent edges, and many earthquakes also occur at these plate boundaries.

Champagne pool, Wai-O-Tapo, New Zealand

THE COLOURFUL, STEAMING CHAMPAGNE POOL in the Wai-O-Tapo (meaning 'sacred water') geothermal reserve of New Zealand, is heated to 74°C / 165°F, and carbon dioxide bubbles pop and burst at the surface. It is so rich in minerals that its waters are stained like an artist's palette: its bright orange rim almost otherworldly. The 65m- / 213ft-wide, 62m- / 203ft-deep pool formed 700 years ago from a volcanic explosion, and is fed by a deep hydrothermal reservoir of magma-heated water containing dissolved gold, mercury, arsenic and silver. The region's geothermal activity dates back over 150,000 years, and is caused by pressure of the Indo-Australian Plate over-lapping the Pacific Plate.

Lake Ngakoro, New Zealand

THE COMBINED EFFECT OF EARTH'S NATURAL WATER and heat create magnificent geothermal gems. For example, New Zealand's North Island, a kinetic part of the Pacific Rim of Fire, has a vivid patchwork of colourful, geothermal lakes. Dissolved sulphur and ferric salts give the 760-year old Lake Ngakoro (Ngakoro means grandfather) its brilliant green hue. The lake is inside a volcanic crater, and was filled by the overflow from nearby pools and streams. New Zealand's geological history of volcanoes, hot pools, geysers, steam vents, geothermal activity and hydrothermal energy have led some to call it 'the shaky isles'.

181

Orakei Korako silica terraces, New Zealand

ORAKEI KORAKO IS AN ISOLATED GEOTHERMAL VALLEY, famous for its showy silica terraces situated along the banks of the Waikato River in the Taupo volcanic region of New Zealand. Dissolved silicate of lime, called sinter, is deposited as water evaporates, gradually building up colourful terraces. They are caused by various types of cyanobacteria, or water algae, in the warm water which creates yellow, olive and rust streaks of colour. As much as 20 million liters / 4.4 million gallons of silica-rich water flow over the terraces every day. Orakei Korako also has more than 30 geysers, and the native Maori people once used these hot springs for cooking and cleaning. Their modern relatives still consider themselves the guardians of these ancient, geothermal features.

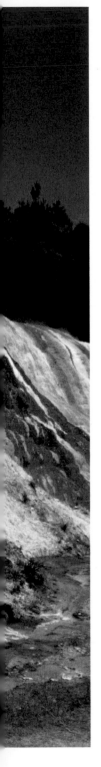

Shield volcano, Piton de la Fournaise, Réunion Island

THE SHAPE OF A VOLCANO tells us what type of lava has erupted. Effusive fluid basalt lavas spill over long distances, building up in thin layers to create broad, gently domed shield volcanoes with graceful, gradual slopes. Piton de la Fournaise volcano, on the French Territory of Réunion Island in the western Indian Ocean, has erupted fluid lava flows more than 150 times since the mid-1600s. The top of the volcano is indented with an 8km- / 4.9 mile-wide collapse caldera. About half a million people live on the French island beside Piton, one of the most active volcanoes on Earth. In 1977 the town of Piton-Sainte-Rose was assaulted by a lava flow causing evacuations and severe damage.

Old Faithful, Wyoming, USA

'OLD FAITHFUL' IN WYOMING'S YELLOWSTONE National Park, USA, is one of the world's most famous geysers. It erupts for an average of four minutes about once every 94 minutes, squirting hot water 45m / 147.6ft in the air. Because its mouth is shaped like a funnel, the water streams straight up in a slim spurt. As with most geysers its eruptive patterns and behaviour can be changed over time by earthquake disturbances, temperature fluctuations, water pressure variations and gas content. For example, in 1998 an earthquake caused Old Faithful's eruption intervals to lengthen. Predicting the exact time a geyser will erupt is difficult, even with Old Faithful, and for decades scientists and even mathematicians have tried to unravel the mystery.

Intra-plate volcanism, Hawaiian island chain hot spot

MANY VOLCANOES FORM AT PLATE BOUNDARIES, but not all; some grow in the middle of a tectonic plate. In the 1960s, scientists studying volcanic island chains in Hawaii noticed that the rocks on the islands became younger towards the southeast. This suggested that each island formed over a stationary hot spot, a plume of magma burning through the Pacific Plate, and then slid northeast with the movement of the plate. This is the theory behind intra-plate volcanoes. At present, the most southern Hawaiian island, the Big Island of Hawaii, sits over the hot spot and has five active volcanoes. Eventually it will move northeast, become extinct, and a new volcanic island will form over the hot spot. Over the past 10 million years more than 100 hot spots have penetrated the Earth's crust.

Super-eruptions from super-volcanoes, Taupo, New Zealand

TAUPO VOLCANO IN NEW ZEALAND is an island super-volcano. Super-volcanoes have such explosive, violent eruptions that entire continents are devastated. So much gas and ash is shot into the atmosphere that sunlight cannot
penetrate, causing the Earth's temperature to cool rapidly, resulting in instant global winter and a mini ice age.
Thick layers of acidic ash are blown over tens of millions of square kilometers, polluting water supplies and causing crops to fail, resulting in massive starvation and millions of deaths. New Zealand, the USA, South America, Asia
and Europe all have super-volcanoes that could erupt with fatal consequences. Scientists believe another super-
eruption at Taupo is inevitable.

Lava Butte, Oregon, USA

LAVA BUTTE NEAR BEND IN OREGON, USA, is composed of volcanic cinders that sprayed into the air from Newberry volcano, then fell back down creating a flat-topped cone 7,000 years ago. A road winds to the summit, 166m / 545ft high, and offers visitors an unparalleled view of the surrounding Cascade Mountains. Over 25 sq km/ 9.4 sq miles of ground was ultimately covered with lava flows from that one eruption. Lava Butte has remained virtually intact because of its porosity, allowing the wind and rain to filter through. More than 400 other cinder cones and lava formations have been found along the fissures of Newberry volcano.

Hornito, Etna, Italy

HORNITO IS A SPANISH WORD meaning 'little oven'. Unlike spatter cones that form over vents hornitos are 'rootless' and have no underlying vents. Instead they form over holes, or perforations, in the tops of crusted lava or over lava tubes. As molten lava gushes through a tube, wads are forced up and ejected out, building a lumpy pile around a perforation. Hornitos have steep sides of stacked clumps, with red lava slopping out the top. If the side of a hornito collapses, then lava can spill out in a flow. Mt. Etna, with its criss-crossing lava tubes which formed during numerous past eruptions, has multiple hornitos, and some have collapsed letting lava escape into the Valle del Bove.

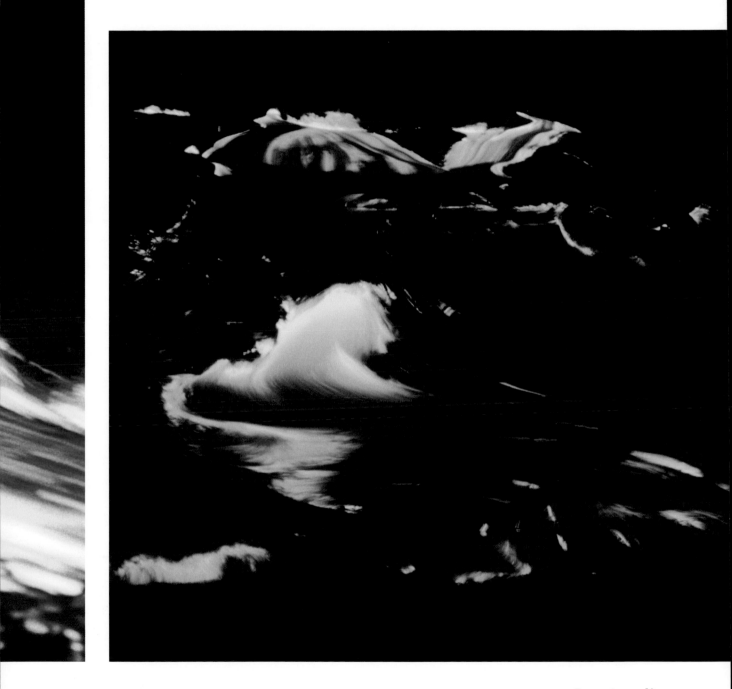

Lava tumuli

WITHOUT AN INCLINE, FLUID BASALT LAVA CREEPS slowly across a flat surface as its crust cools. As it inches along
the lava pools in natural depressions or is blocked and backed up against obstacles, then cools, forming a crust.
Pressure builds up as gasses in the hot lava expand, inflate and upwell under the brittle crust. Eventually the hot
lava balloons up in big rounded heaps, up to 10m / 33ft high, called tumuli. The fragile crust buckles up with it and
snaps open at the top, allowing lava to gush out of the split. Kilauea volcano in Hawaii has hundreds of bulging
tumuli on its tremendous lava flows. Tumuli is the plural of the Latin tumulus, meaning a mound or hill over a grave.

Volcanic ice cave, Iceland

WHEN COLD GLACIERS AND HOT VOLCANOES MEET, glistening blue ice caves can form. Volcanic ice caves are shaped by sublimation, when solid ice is scorched instantaneously into vapour without passing through a liquid phase, as glaciers encounter hot steaming fumaroles or magma-heated rocks. Once carved, circulating geothermal heat keeps ice caves hollow and stable for long periods. In this scenario the floors of the caverns are usually warmer than the walls. Some minor melting can occur, and stalactites of ice and crystals melt and refreeze. Iceland, with its multiple volcanoes and glaciers, presents the perfect conditions for the formation of volcanic ice caves.

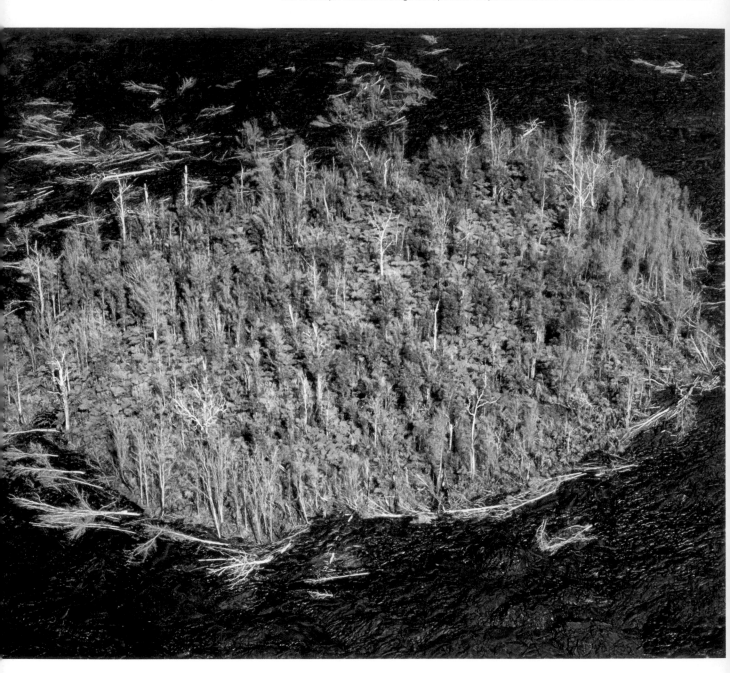

Kipuka, Hawaii, USA

190   KIPUKA IS THE HAWAIIAN FOR ISLAND. When fluid basalt lavas inundate entire forests of trees, occasionally an island of green trees on a hill, mound or rise remains intact as the lava circles around it. This island of green trees is called a volcanic kipuka and in time new plants grow around the kipuka that's standing above the black, surrounding lava flow. Meanwhile these old, fragile forest kipukas become extraordinary eco-systems from the past, and they are of great interest to biologists because they contain rare mature specimens of native species of plants, birds and animals, otherwise wiped out by the lava flows. In Hawaii, these ecological treasures are being preserved and protected.

Lava trees, Lava Tree Park, Hawaii, USA

FLUID BASALT PAHOEHOE LAVA FLOWS FLOOD forests of trees, burn the trunks and preserve their shapes in a solid-ified, lava tree cast or 'lava tree'. The growing trees are much cooler than the hot lava that rushes up to meet them, so when it makes contact the lava chills and then hardens against the bark, before the tree burns. When the flow drains away all that's left is a tall, hollow tube of vertical lava in the tree's mould. Sometimes the inside of the lava tree has indentations from the parent tree's bark texture. Lava Tree State Park in Hawaii, USA, contains the rem-nants of a lush Ohia tree forest that was deluged by a hot lava flow hundreds of years ago.

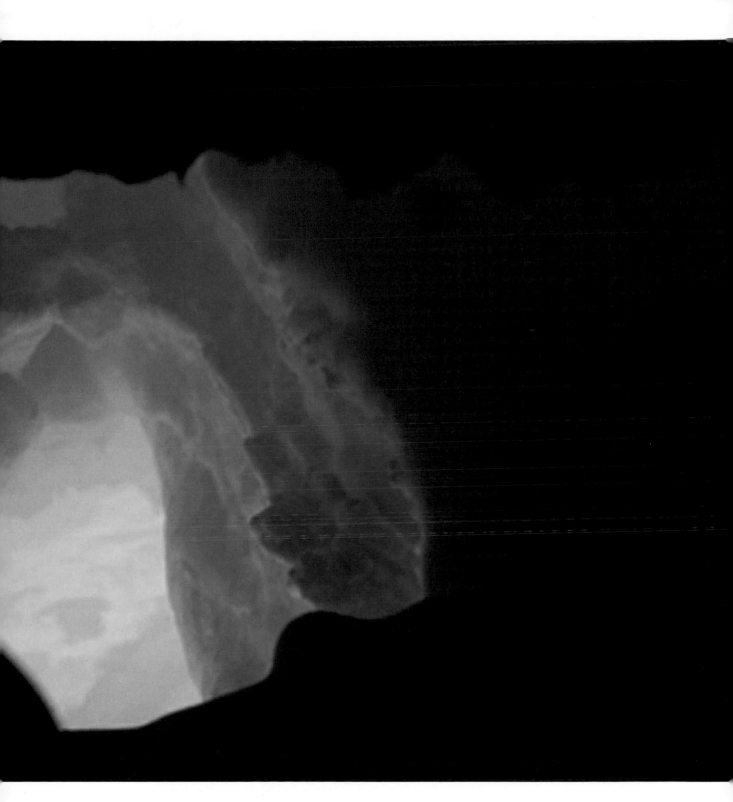

Lava tube skylight, Hawaii, USA

THE BIG ISLAND OF HAWAII is remarkable for its fluid, basalt, lava rivers that cool and harden on the top and sides, forming sealed tubes with red-hot lava flowing inside them. Because it is insulated in the tubes, the molten lava can flow great distances, as much as 10km / 6.2 miles. On rare occasions, the top crust or ceiling of the blackened lava tube breaks off, and a hole or breach occurs, allowing us to see into the fiery molten interior. Sometimes these holes crust over and close up once again, but once the eruption ends and lava drains from the tube, we are often left with a cave pocked by skylights. These perforations in the ceilings of lava tubes are called lava skylights.

Volcanic deformation, sea floor spreading, Gulf of Aden, Red Sea

AS CONVECTION CURRENTS IN THE EARTH'S PLIABLE MANTLE swirl in eddies, tension causes the overlying crust to
fracture. Great valleys are stretched apart and flooded with seawater, forming basins and gulfs. The Gulf of Aden
in the Red Sea straddles a spreading, sea floor rift valley along a giant fault near the southern edge of the Arabian
Plate, that formed as the Arabian Plate moved northeast from Africa. Scientists have used submarines to research
this seismic underwater terrain, and a series of seismographs continuously monitor the ocean bottom. If this
spreading persists, Eastern Africa could one day flood.

Solidified lahar path, Mt. St. Helens, Washington, USA

MT. ST. HELENS UNLEASHED DESTRUCTIVE LAHARS (volcanic mudflows) in its deadly 1980 blast. As steam blasted through the summit ice cap and caused the largest landslide in history, volcanic ash and debris mixed with the meltwater creating a hot, paste like deluge. The lahars poured down valleys and river beds at up to 110km / 68 miles per hour, and razed everything in their wake, including buildings, trees, roads, 27 bridges and 200 homes. Once the eruption ended, the lahars dried out leaving snaking channels as dense as concrete. The most destructive lahars surged down the North Fork Toutle River (pictured), ploughing 9km / 5.5 miles downstream. Mt. St. Helens could burst to life again anytime.

Vast geothermal area, Rotorua region, New Zealand

THE ROTORUA REGION OF NEW ZEALAND is a highly geothermal area that is a result of the ongoing collision of the Pacific and Australian Plates. Originally the islands of New Zealand were part of an ancient super-continent called Gondwana, and rocks found in New Zealand are more than 500 million years old. Primitive uplift, fault movements, pressure and melting formed the two mountainous islands and, as the plates continued to subside, volcanism increased. Earth's molten power is reflected in the many various geothermal features at Rotorua: surging geysers, bubbling mud pools, pink mineral terraces, mineral-stained hot ponds and roaring volcanoes. An area called the Artist's Palette (pictured) glitters with vibrant mineral colours.

198

Coral atolls

MOST CORAL ATOLLS – CIRCULAR CORAL ISLANDS WITH SALTY LAGOONS in their centers – begin life as volcanic islands in tropical oceans and Charles Darwin was one of the first scientists to correctly describe coral atolls. They begin life as a pointed volcanic nose sticking out of the water around which tropical marine animals called corals, build calcareous circular reefs. Once the underlying volcano is extinct, wind and waves erode the cone and, eventually, the volcano sinks below water level. Just the enclosed ring of coral remains, which continues to build and grow. The Pacific Ocean, with its abundant volcanoes and its tropical and sub-tropical temperatures that coral need to survive, has the most coral atolls of any ocean. In the Indian Ocean you'll find the Maldives, a group of 26 coral atolls and other islands.

La Perouse Rock, French Frigate Shoal, Hawaii, USA

ABOUT 805KM / 500 MILES NORTH OF HONOLULU, Hawaii, lies the crescent reef of French Frigate Shoal. In the centre of its moon-shaped reef lies La Perouse Rock, a volcanic remnant 27m / 88.5ft thick and 41m / 134.5ft tall. It is believed that long ago a huge volcano, 9km / 5.5 miles in diameter, once stood where La Perouse now docs. When the volcano became extinct and sunk, solidified lava in its main conduit remained standing. Sand and marine debris caught in the resulting reef have created more than a dozen shoals or small islets. The lee side of the curved reef holds a calm lagoon which has helped preserve the pinnacle of volcanic, olivine, basalt rock near its centre.

Green sand beach, Papakolea, South Point, Hawaii, USA

BEACHES COME IN A VARIETY OF COLOURS. Those made of crushed shell can be pink and white, crushed lava beaches are black, and rare beaches made from volcanic, olivine crystals are green. One of the most famous green sand beaches is Papakolea, at the base of Mauna Loa volcano in K'au, Hawaii. The beach is in the breached side of a volcanic, olivine tuff ring called Puu Mahana that blasted up long ago when olivine-rich magma encountered shallow groundwater. Wave action eroded the tuff ring, washed away lighter particles of ash, and pounded the olivine crystals into a pretty bay of green sand.

Giant's Causeway, North Antrim, Ireland

IRELAND'S GIANT'S CAUSEWAY consists of spectacular, hexagonal, basalt lava columns formed about 65 million years ago during the Tertiary Period when the hot, molten rock crystalized and contracted, creating smooth cylinders as it encountered cold water. The result is more than 40,000 tightly clustered, symmetrical columns of basalt stretching 180m / 590ft wide and 150m / 492ft into the rugged sea. The Causeway is named after a small, mythical giant, Finn MacCool who, legend has it, made all this himself.

Pamukkale terraces, Turkey

TURKEY'S EXQUISITE, NATURAL TERRACES AT PAMUKKALE – 2,700m / 8,858ft long and 160m / 525ft high – have no equal. The pale terraces are perched like a bleached smear along the edge of a mountain plateau 100m / 328ft high in the Meander River valley, near the town of Denizli. Geothermally heated mineral springs, rich in calcites, streaming down the tiered hillside have deposited mounds of calcium carbonate and thick, white limestone that solidified into solid, travertine drifts forming tiers. The ivory coloured, scalloped, shallow pools of water and solidified, blanched waterfalls look like puffs of fluffy cotton. No wonder the locals call this 'The Cotton Castle'. The ancient city of Hieropolis is built on top of Pamukkale.

Volcanic sulphur deposits

BRITTLE, LEMON-YELLOW SULPHUR NATURALLY COOLS into solid crystals after it has erupted from hot, gassy fumaroles on volcanoes. Sulphur is mined and used to make gunpowder, rubber, paper, plant fertilizers, wine preservatives, matches, batteries, fireworks and insecticides, and also makes good electrical insulation. More importantly, human cells need sulphur to survive, and a portion of human body fats, fluids and skeletal material contain sulphur. When combined with oxygen, a dangerous polluting compound called sulphur dioxide is produced. Hydrogen sulphide, which smells like rotten eggs, can even cause death by asphyxiation, and when melted turns vibrant red. Sulphur deposited and mined from Usu volcano near Hokkaido, Japan, boosts the local economy, as it does elsewhere.

Eruption and collapse, Waimea Canyon, Kau'ai, Hawaii, USA

LIKE THE OTHER HAWAIIAN ISLANDS, Kau'ai is the ancient tip of a volcano emerging from the Pacific. Kau'ai's elaborate Waimea Canyon – 10km / 6.2 miles long and 900m / 2,952.7ft deep – is a spectacular series of colourful, interwoven gorges forged by erosion and gargantuan collapse. Kau'ai is one of the rainiest places on Earth, and the powerful Waimea River – fed by rain-laden swamps – has carved through layers of basalt over the centuries. Wind and weather have coloured the exposed layers red and tawny yellow and the western region of Waimea Canyon is layered with thin lavas while the east reveals thicker flows. The two areas are separated by a massive fault, along which a section of the island slipped and dropped 4 million years ago. Mark Twain named it the 'Grand Canyon of the Pacific'.

Volcanic soil, Atacama Desert, Chile

THE ATACAMA DESERT IN CHILE is one of the driest places on Earth, with vast salt flats and ancient lava flows stretching from the Andes Mountains to the Pacific Ocean, covering 181,000sq km / 69,884sq miles. The desert is walled-in on both sides by mountain ranges containing towering, snow-capped volcanoes whose ancient lavas have eroded into brown, volcanic soil. The majestic Andes block virtually all precipitation from reaching this desert, which means it has an approximate average rainfall of just 1mm / 0.04in per annum. In fact centuries have passed with no rain at all. Geologists have compared the barren soil here to that on Mars, and NASA uses the region to test space equipment.

Hot springs, Jigokudani, Japan

THE RED-FACED MACAQUE MONKEYS of remote Jigokudani, in central Japan, enjoy bathing in the natural volcanic hot springs and sulphur ponds to beat the cold and snow. Geothermally heated water, about 50°C / 122°F, gurgles and steams out of the frozen cracks and fills ponds in the snowy landscape, attracting these 'snow monkeys', who visit the hot springs only during the winter months. Like other primates, Japanese macaques have a highly regimented social and family hierarchy, and visitors are not allowed to bathe with them. A ski resort in the vicinity does have a hot bath exclusively for humans, and occasionally the monkeys dive in. The region's nickname is Hell's Valley.

Moss grows on Antarctic fumerole

IN THE FREEZING ARID CLIMATE OF ANTARCTICA, opportunistic mosses take advantage of the warmth seeping from steamy fumarole cracks on volcanoes. Mosses, members of the family of Bryophytes, don't produce flowers or seeds but clump together in green mounds of tiny leaves and stems. They take advantage of these unlikely ecological niches, which provide a temperate climate, a constant water source from condensed steam and nutritious minerals leached from surrounding, volcanic rocks. Over 100 types of mosses can be found in Antarctica, many growing on volcanic fumarole sites.

Volcanic rock formations, Cappadocia, Turkey

HISTORIC VOLCANIC ERUPTIONS FORMED A NATURAL wonderland in Cappadocia, Turkey and one which was subsequently inhabited. Ancient civilizations tunneled through 500m / 1,640ft of thick, soft layers of volcanic ash and mud called tuff, digging out thousands of subterranean and cliff-hanging cave towns in which to hide from would-be conquerors. Some of these primitive dwellings are still inhabited, where the locals follow a simple way of life. One of the earliest cave paintings anywhere was found here, and depicts a volcano erupting. Bizarre rock formations, called fairy chimneys, created by wind and water erosion, dot the terrestrial landscape

Destruction of landscape, Piton de la Fournaise, Réunion Island

LAVA FLOWS, WHETHER RUNNY PAHOEHOE or tumbling, rubble-packed aa, are lethal. Homes and buildings burn, trees are buried, and roads enshrouded for miles. In 2001, Piton de la Fournaise, a basalt shield volcano over 500,000 years old and one of Earth's most active volcanoes (with over 170 eruptions since the 1700s) erupted forceful aa flows that knocked over forests, melted asphalt and covered a road between the eastern and southern part of Réunion Island before reaching the ocean. Over half a million people living on the island are used to its blasts, fireworks and frequent lava flows.

VOLCANIC LAND FORMS

Alo'i Lava Cascades, Mauna Ulu, Hawaii

BLAZING LAVA CASCADES flow downstream and over cliff faces, just like rivers of water. During the 1969-71 eruption of Mauna Ulu, a flank vent on Kilauea volcano, Hawaii, raging cascades dazzled onlookers as they plunged 30m / 98ft off a cliff precipice. These flows actually began as 540m- / 1,771ft-high lava fountains near Alo'i pit crater, spilling down the slope and then plummeting over the cliffs, changing the landscape and adding new land to the island. The speed of the tumbling flows changed the lava from basalt pahoehoe to aa, and several sections of the highway, the Chain of Craters Road, were buried under lava.

Lava flames eating vegetation

THE COLOR OF A LAVA FLOW can indicate its approximate temperature. The coolest lava flows are a deep dull red at around 650ºC / 1200ºF. As temperatures increase the lava color changes from deep red to orange, next to yellow-orange and finally the hottest, yellow, at around 1,050ºC / 2100ºF. The top of a lava flow cools instantly into a silvery gray then black crust and the lava within stays red hot and glowing. The only way scientists can determine the exact temperature of a lava flow is by inserting a lava thermometer called a thermocouple probe.

VOLCANOES CREATE SOME OF THE EARTH'S NEWEST LAND, ADDING IT IN TWO WAYS. Firstly, lava flows build up and up in horizontal layers on solid earth; as well as this, when the rate of lava flowing into the sea is greater than the rate of wave erosion, new land forms, expanding the coast at the shores of islands, and the edges of entire continents. No one knows for sure exactly how much new volcanic land is added to the Earth each year but geologists estimate it is approximately 42 million cubic meters.

When really huge eruptions occur, much more land is manufactured, disproving the preconception that volcanoes are entirely destructive forces. Over 60,000 kilometers / 70,000 miles of the Earth's deep ocean floor is covered with active volcanic ridges formed by submarine rift volcanism, such as occurs at the Mid Atlantic Ridge. This is a vast area, and about one third of all of Earth's recorded volcanoes pop up on on mid-oceanic islands and in island arcs. Iceland, in the Atlantic Ocean, is one of the only places on Earth where a volcanic undersea ridge actually breaks the surface of the water, resulting in spectacular volcanism. No one on Earth knows exactly how many submarine volcanoes exist but there may be as many as one million. Undersea or submarine volcanic eruptions are the most common eruptions on Earth. Volcanoes and the sea continually battle with each other in many ways. The weight of water (in volcanic lakes and under the sea) upon lava flows produces unusual balloon or bulbous-shaped lava flows named pillow lava.

When terrestrial volcanoes located on solid ground pour hot lava into the cold sea there are many dramatic interactions, and often explosions occur. The results of hot lava meeting water are often beautiful, but also lethal. Salt water can be intensely cooked by hot lava, producing hazardous steam clouds containing hydrochloric acid and when volcanoes release sulfur dioxide into the air, the gas combines with atmospheric water vapor to form acid rain, that can devastate vegetation, corrode building materials, and cause respiratory ailments. Violent steam blasts of lava and water occur when hot molten lava meets the cold sea such as at Stromboli volcano in Italy. Modern technological innovations and deep sea equipment have recently enabled us to virtually witness undersea volcanoes erupting. On occasion, these undersea volcanoes become land masses themselves: water, steam, lava and cinders are erupted until they finally burst free as a new island. Surtsey Island in Iceland formed this way between 1963 and 1967.

VOLCANOES & THE SEA

5

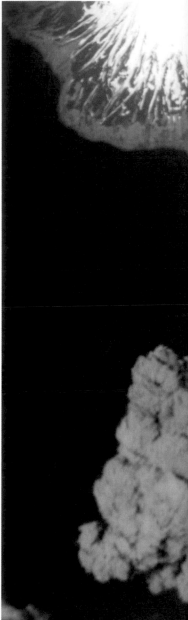

Subducted volcanic islands, Saipan, Northern Mariana Islands

THE EDGES OF THE EARTH'S TECTONIC PLATES are the birth-place of about 95 per cent of the planet's volcanoes. Where the Pacific Plate is being pushed under the Philippine Plate beneath the Pacific Ocean, the melted rock oozes out along the seam. That escaped lava has built a crescent-shaped ridge of strato-volcanoes In Micronesia called the Mariana Islands. One of the most populated, Saipan, has spectacular cliffs that are the remains of an ancient steep volcanic cone. In 2003 Anatahan volcano, just north of Saipan, suddenly and unexpectedly burst to life blasting ash 10 km / 6.2 miles in to the air. A typhoon blew the toxic cloud over Saipan, drenching the island in foul smelling ash fall.

Previous pages | Steam blast, lava meets sea water

RUNNY HOT BASALT lava pours down the flanks of Kilauea, a Hawaiian shield volcano, until it encounters the cool surrounding waves of the Pacific Ocean. Tremendous plumes of steam are formed and explosions occur as the lava instantly boils and vaporizes the ocean water.

Volcanic island chain, Aleutian Islands, Alaska, USA

ALASKA'S RUGGED ALEUTIAN ISLANDS, stretching 1,900 km / 1,180 miles, are one of the longest chains of volcanic islands in the world. Fed by a constant flow of magma, they were created as the Pacific Plate is pushed beneath the North American Plate. Fifty-seven active volcanoes are sprinkled over more than 300 islands that form a fence between the Bering Sea and the Pacific Ocean. One of the most active Aleutian volcanoes is Mt. Cleveland, a steep (1,730m / 5, 676ft) pointy strato-volcano, on Chuginadak Island. In 2006 Mt. Cleveland spewed a dark brown column of ash that was photographed by the International Space Station (above). The Aleutians are part of the Pacific Ring of Fire and Chuginadak is the Aleut goddess of fire.

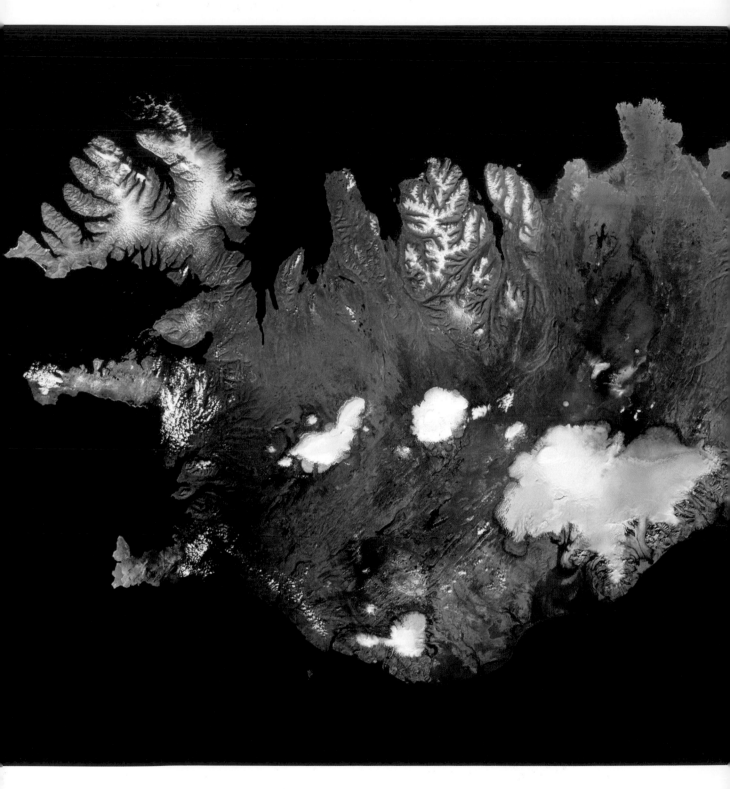

Largest volcanic island, Iceland

ICELAND, BORDERING THE ARCTIC CIRCLE in the North Atlantic Ocean, is the largest volcanic island in the world. The Mid-Atlantic Ridge dissects the island making it highly geologically active. Its active volcanoes include Hekla, Laki and Eldfell. In the 1700's ash eruptions from Laki sent so much volcanic debris into the atmosphere that crops failed and famine followed. Since 10 per cent of Iceland, and its hot active volcanoes are covered in glaciers, massive floods of geothermally-melted glacier water, called jokulhaups, often deluge the island. Icelanders have harnessed the island's bounty of volcanic processes for geothermal power, and benefit by heating their homes and water for very little cost.

World's Northernmost volcano, Beerenberg, Jan Mayan

THE STRATO-VOLCANO BEERENBERG IS ATTACHED to the tip of Jan Mayan island, in the North Atlantic ocean, by a thin isthmus of land. Sitting atop a fracture zone it is fueled by magma produced in the Mohns mid-ocean ridge and the volcano has erupted about 6 times in the last 200 years, mostly from flank vents. A 1970 eruption added 3 km / 1.8 miles of new land to the island and in 1985 a fissure cracked open and erupted 7 million cubic meters of lava in less than 40 hours. So much lava entered the sea that the temperature of the water temporarily rose from a normal of 0°C / 32°F to about 30°C / 86°F.

Amutka, Aleutian Islands, Alaska, USA

AMUTKA IS AN UNINHABITED STRATO-VOLCANO in Alaska's Aleutian island arc. No one knew much about it until a pilot flying near the volcano in 1987 noticed a huge column of dark ash rising above the clouds. The massive plume drifted 16km / 10 miles to the south. The circular volcano, 7.7 km / 4.7 miles wide, encompasses the entire island and overlies an ancient broad shield volcano. Like other volcanoes in the Aleutian chain, Amukta is part of the Pacific Ring of Fire, and is fed by lava formed as the Pacific Plate is crushed and melted under the North American Plate. Amukta has had about five big eruptions since the mid-1700's.

Submarine volcano, East Pacific Rise ▽

SCIENTISTS ESTIMATE there are more than one million submarine volcanoes, producing 75 per cent of the Earth's magma. Submarine volcanoes add new crust to the sea floor and at the edges of tectonic plates. In 2005 GLORIA (Geologic Long-Range Inclined Asdic), a long-range, sonar-scanner, discovered a new erupting submarine volcano near the East Pacific Rise, west of Mexico. This region that is experiencing sea floor spreading, when two of Earth's plates slide apart, allowing magma to rise and erupt on the sea floor.

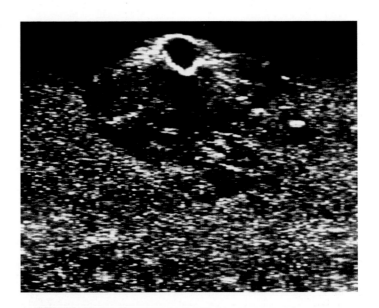

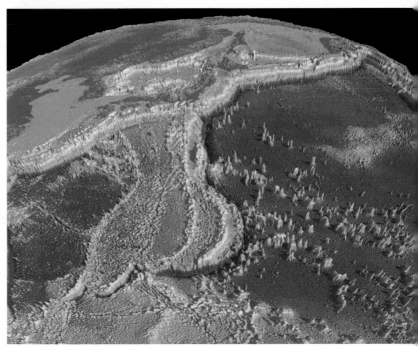

The Marianas Trench △

THE MARIANAS TRENCH, a vast crevass that formed along the edge of the Pacific Plate and Philippine Plate, is the deepest point on earth and dips at a steep angle to about 11,000m / 36,089ft inside the Earth. As a descending plate melts into buoyant magma it rises to feed and build island arcs of volcanoes, in this case the Mariana Islands. In this computer image depth has been color coded: red is the highest, then yellow, green, blue and purple the deepest. Land is dark green. The Marianas trench is pictured as a purple arc.

221

Kurile Islands, Russia

RUSSIA'S KURILE ISLANDS ARE A CHAIN of more than 150 volcanoes that span 1,250 km / 777 miles from Kamchatka to Japan. The emergent volcanic archipelago is the surface spine of an intensely volcanic area called The Greater Kurile Ridge which rises from the bottom of the sea. The Kurile volcanoes formed about 90 million years ago as the Kurile Plate was forced beneath the Siberian continent. Next the Kurile Plate migrated south, pushing the Pacific Plate under its southern edge, adding to the region's volcanic drama. At any given time about 40 Kurile volcanoes are active. The largest eruption in the history of the Kuriles occurred at Alaid volcano in 1981.

Submerged volcano, Wake Island Atoll

WAKE ISLAND RESTS IN THE PACIFIC about halfway between Hawaii and Guam. The three islands that comprise Wake: Peale, Wake and Wilkes are coral atolls that grew along the rim of an extinct and collapsed volcanic crater in the shape of a turkey wishbone. After the volcano sunk the coral continued to grow and a tranquil turquoise lagoon now fills the depression where the volcano once stood. Over time, waves and wind have eroded the island's coral edges into a ring of soft white sand beaches. Wake Island became famous during the Japanese attack on Pearl Harbor, Hawaii, when marines stationed on Wake Island held their ground against a massive attack.

Slope gradient, lava speeds towards sea, Kilauea, Hawaii, USA

AS LAVA ERUPTS ITS VISCOSITY, magnitude, rate of cooling, and the gradient of the slope upon which it is erupted, govern how fast and far it will flow. Its viscosity, (how thick or thin it is) depends on its silica content (the more silica the denser and slower) and limits whether it travels 100 km or 1 metre. The more fluid and runny the lava the further it flows before cooling. Total volumetric output and force is also a confining factor – if lava dribbles out of a vent it won't go far before getting cold and solid. Finally the geology of the volcanic terrain is important. Steep slopes equal acceleration. Flows build narrow channels and tubes, which mean lava stays hot and fluid and travels faster and farther. In Hawaii effusive silicon-poor, basalt lava can travel 12km / 7.4 miles uphill from a source vent, before pouring into the Pacific Ocean.

Creation and collapse of lava bench, Hawaii, USA

WHEN PAHOEHOE LAVA FLOWS TO THE COAST and enters the sea an amazing transformation occurs. New land instantly forms as it cools into a fan-shaped rock shelf called a lava bench or delta. This lava delta, only a few days old, extended the island of Hawaii about 150m / 493ft with the newest land on Earth. As waves pounded the lava delta it developed dangerous cracks and the leading edge of the lava bench, seaward of these cracks, became unstable, weakened and then caved into the sea. As long as the supply of lava continues a lava delta will repeatedly solidify, grow in width and length and have its leading edge collapse. When the rate of lava is greater than the rate of wave erosion then the island slowly advances into the sea with new land.

Previous pages | Battle between lava and sea

SEVENTY PER CENT OF EARTH IS COVERED IN OCEAN and Earth's tides rise and fall twice a day. This up and down movement covers and exposes already hardened coastal lava flows. It also affects active lava flowing into the sea. For example, lava pouring from a tube into cold sea water may seal off the open end at high tide. As tides recede, waves pound the thin covering, reopening it. When lava builds shelf-like benches out into the sea, big waves smash the bench exposing hot flows. Lastly whether a new unstable volcanic island, a new lava shelf or delta, or an active flow, if the ocean waves are strong enough then the new land is washed away. The battle between the sea and volcanoes rages all round the world.

Underwater volcano, Iwo Jima, Japan

IWO JIMA (JAPANESE FOR SULPHUR ISLAND) IS PART OF a chain of islands off Japan called the Volcano Islands. In July of 2005, Japanese troops stationed on Iwo Jima spotted a steaming plume 1100m / 3,609ft tall, rising from the Pacific ocean about 48km / 30 miles in the distance. Coast Guard crews rushed to the site and found a big patch of bubbling, muddy gray and red water. They realized it was a shallow undersea volcano erupting. This area, called Fukutokuoka-no-ba, had similar underwater volcanic eruptions in 1986 that lasted three days and as future eruptions build up this new underwater volcano's edifice it may yet emerge above the ocean's surface as a new volcanic island.

Sciara del Fuoco eruptions and tsunami, Stromboli, Italy

STROMBOLI VOLCANO IN ITALY IS FAMOUS for its jetting fountains of lava but occasionally it has effusive outpourings. The side of Stromboli has collapsed repeatedly leaving a long black indented scar called the Sciara del Fuoco (Road of Fire) about 900m / 2,953ft long. Effusive flows burn their way down the Sciara del Fuoco as it offers the route of least resistance. In 2002 a lava flow 300m / 984.3ft wide crashed down the Sciara del Fuoco and caused a mighty avalanche into the ocean. Two subsequent tsunamis followed, spraying the island with scalding water that damaged buildings, boats and injured bystanders. Large waves were recorded 60km/ 37.3 miles away on the island of Sicily.

WHEN HOT (1,100°C / 2,012°F) LAVA RUSHES into the sea stunning tephra jets – caused by intense explosions – shoot radiant chunks of lava sailing up through the air, and cold sea water instantly flashes to volatile steam. At sea coasts where a lava delta – a hardened lava shelf that extends into the sea – has formed, a total collapse of the delta's forward edge can expose lava to come into contact with cold sea water, producing massive tephra jet explosions and blasting hardened chunks of the delta as much as 200m / 656ft inland.

Toxic steam plume

AS LAVA ENTERS THE OCEAN tall columns of hot toxic steam plumes form as the water is instantly vaporized, the lava explodes and the surrounding ocean heats up. The water is about 69°C / 156.2°F at the lava entry point and ranges to about 35°C / 95°F at 100m / 328ft from the source, and the size of the steam plume is dependent on the amount of lava entering the ocean. Lava steam plumes are hazardous: the surrounding scalding water can splash far inland and chloride from sea salt combines with hydrogen saturating the plume with toxic hydrogen chloride. When offshore winds blow this mixture inland it can produce a disorienting white-out cloud on the ground, and also often falls as a corrosive acid rain. Tiny shards of glass, formed when the lava cools, are also carried by the plume and can penetrate skin, nasal passes and eyes when they fall.

Green trail from underwater eruption, Iwo Jima, Japan

LIKE PEOPLE, VOLCANOES ALSO HAVE life stages. They are born, they live and they die. In 2005 a new shallow submarine, or underwater, volcano near Iwo Jima burst to life, erupting hot sulfur and ash just below the ocean's surface. Ocean currents pulled the colorful volcanic debris across the water in a long yellow trail. Ships were warned to stay away from the hydro-explosions as sea water was vaporized into a hot tall column of steam. Shallow submarine eruptions offer scientists a remarkable opportunity to study how undersea volcanoes work. It is probable that this new volcano has been building to its current height over decades through repeated outpourings of lava and other volcanic debris.

Home Reef volcano, Tonga

IN 2006 A SAILOR NOTICED WHAT APPEARED TO BE HUGE RAFTS (3 km / 1.8 miles wide) of brown styrofoam floating in the ocean near Tonga. Investigators discovered they were actually massive blocks of volcanic pumice erupted from a new island-building volcano named Home Reef. Pumice is a form of volcanic glass saturated with so many bubbles it looks and floats like a sponge. Scientists on-scene reported multiple steam plumes and explosions where lava broke the ocean's surface in four points all issuing from a central crater. Home Reef volcano has built ephemeral islands in the past. In the mid-19th century and in 1984 new islands formed over the volcano and then were eroded away by strong ocean waves. The current island is about 800m / 1,625ft long with a hot water lake in its center.

Does the Moon influence volcanic eruptions?

SCIENTISTS THINK THE GRAVITATIONAL PULL of the Moon and Sun in Space influences volcanoes here on Earth. Two tidal bulges – when the sea literally protrudes out like a dome – occur daily due to the Moon and Sun's gravitational tug of war. As the ocean's tides bulge, the force also squeezes out the Earth's crust. The strongest pull is at New Moon and Full Moon (when the tides become more pronounced) as the sea and the Earth's crust gravitate towards the Moon. Whether the tug at the Earth's crust is enough to trigger or increase volcanic eruptions remains up to science to confirm. If research proves a definite interaction then phases of the Moon could be employed to help forecast volcanic eruptions and save people's lives.

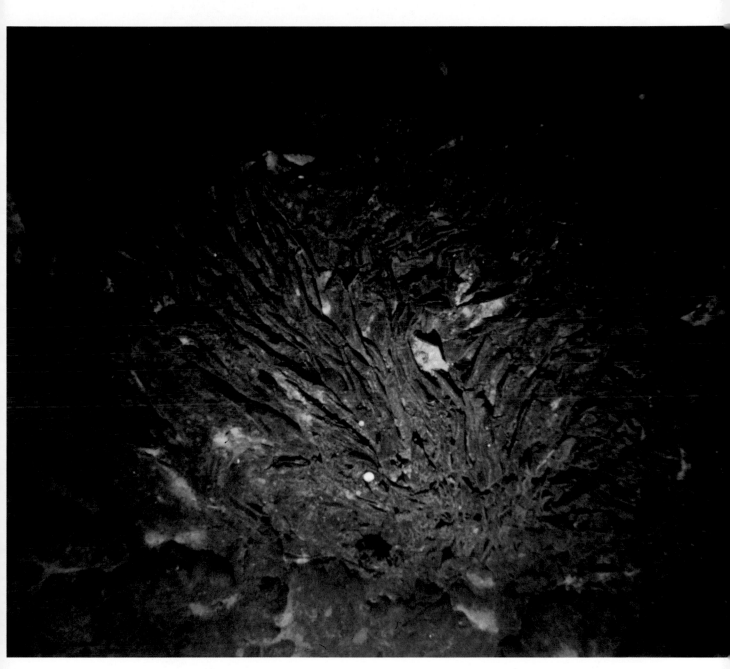

Undersea sheet lava flows

THE MAJORITY OF EARTH'S VOLCANIC ERUPTIONS go undetected because they happen at the bottom of the oceans. Modern technological advances have now given us tools to look beneath the sea and study submerged lava eruptions, revealing the Mid Atlantic Ridge had high-volume, eruptions of fast, runny undersea sheet flow lavas. The crust of these wide flows cools and hardens into a jumble of surface crust shapes and textures. Bulging 'lobate' flows, similar to pahoehoe on land, form when sheet flows progress forward one smooth 'toe' at a time and if molten lava flows out from under a brittle crust the top becomes scratched with lines, cooling as a lineated sheet flow. When fat sheet flows grow thin malleable crusts, the surface is pulled along and folded into curls and coils called ropey sheet flows.

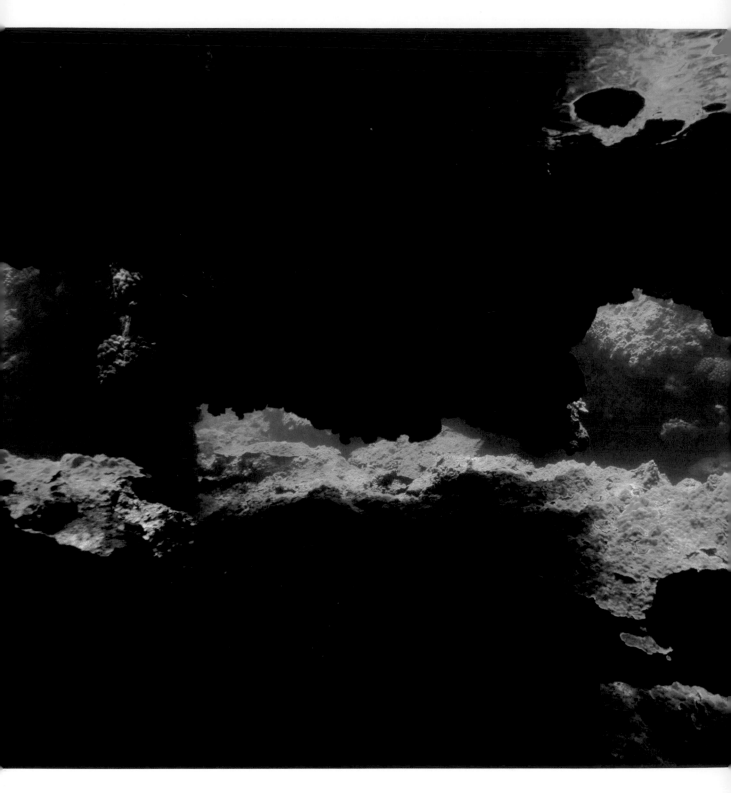

Coral colonization of underwater lava

236 THE HAWAIIAN ISLANDS FORMED OVER A HOT SPOT, a stationary plume of magma from deep within the Earth, that
burned through the crust. Over hundreds of thousands of years living corals grew along the coastlines of each
island. The largest coral reef (232,000 acres / 938sq km) in the Northern Hawaiian Islands colonized the top of old
lava flows. As the Pacific Plate moved northwesterly it dragged each island and its clinging reef with it until, further
away from the hot spots, the volcanoes became extinct and sank into the ocean. The living coral reefs continued
to grow and build outwards in circular and crescent shapes called atolls. In 2006 this coral reef was designated
part of the Northwestern Hawaiian Islands National Monument.

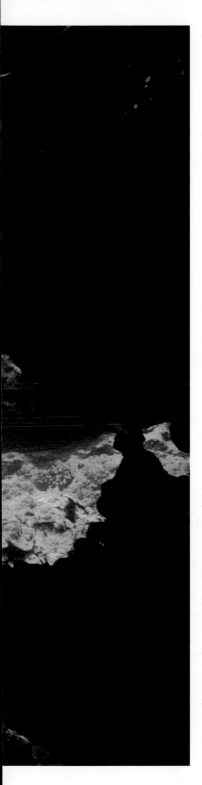

Underwater pillow lava

AS HOT FLUID BASALT LAVA FLOWS INTO THE SEA or oozes from underwater vents, its outer layer freezes into a flexible glassy crust that insulates the molten interior. Gasses within the molten lava swell and inflate the lava forcing it to bulge out in sphere shaped lobes, like toothpaste squeezed from a tube, that pile up in bulbous blobs and mounds called pillow lava. Upon cooling, pillow lava lobes are solid all the way through. Pillow lava is common as the majority of the Earth's volcanoes are located under the sea.

Giant tube worms colonize hydrothermal vents and lava

GIANT (2.4 M / 7.8FT) RED AND WHITE TUBE WORMS (*Riftia pachyptila*) thrive over 1.6 km / 0.9 miles under the sea, congregating around hot, sulfur-rich, volcanic, hydrothermal vents. The top of the worm looks like a frilly red feather swaying in the currents as it gathers food and expels waste. They do not have a digestive system; rather they have a symbiotic (mutually dependent) relationship with the bacteria that live inside their bodies that turn carbon dioxide, oxygen and even hydrogen sulfide into organic nutrients that the worm absorbs and it is believed they evolved this way due to the absence of sunlight. The bottom part of the worm is enclosed in a thick, white tube of skin where the festive plume disappears when under threat.

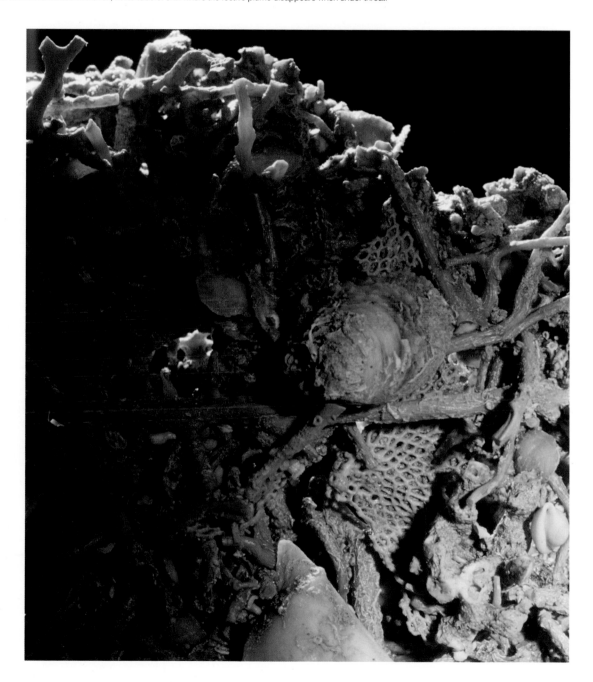

Ferdinandea Seamount, Sicily, Italy, coral study

FERDINANDEA, AN UNDERWATER VOLCANIC SEAMOUNT, lies on the volcanically active, Mediterranean seabed south of Sicily, Italy. Coral samples from Ferdinandea have been salvaged and studied and they show that past eruptions of lava have built up Ferdinandea to above sea level about half a dozen times. Today a large volcanic eruption from Ferdinandea could cause a devastating tsunami and, indeed, in 2002 Ferdinandea registered new seismic activity. In 1831 The King of Naples, Ferdinand II, sent a vessel and christened the new volcanic island Ferdinandea to prevent others from claiming it but by 1832 ocean waves had eroded it 8m / 26ft below sea level once again. Italy's government then sent divers to place an Italian flag in its summit. Ferdinandea is once again about 6m / 19ft below sea level.

DEEP SEA HYDROTHERMAL (HOT WATER) VENTS, that emit what look like champagne bubbles, were discovered 1200m / 400ft off the coast of Papua New Guinea by a team of scientists. These hydrothermal vents, that spew tall columns of hot mineral laden water, are actually underwater geysers. Sea water seeps in, gets heated, then violently blasts out of sea floor vents. Sites around these vents have been found teeming with life attracted by the erupted chemicals and minerals. Vent communities include tube worms, crabs, giant clams, and blind shrimp. All the vents thus far discovered have been found near regions where the sea floor crust spreads, thins and cracks – in other words along mid-oceanic ridges.

Black smoker chimney vents

THE DEEP SEA HAS MID-OCEAN RIDGES where the Earth's crust splits apart and lava erupts. Sections of these volcanic ridges have vast undersea vent fields stacked with 'black smoker' chimney vents. Like geysers, black smokers spew superheated (400°C / 750°F) mineral-rich water but they get their name from the sooty color of the mineral particles that they also erupt. As the hot water cools in the surrounding ocean the minerals: sulfide, metals and silica, deposit tiny fragments and build huge, stiff chimneys, weighing several tons, around each belching vent. The first black smoker vent field was found in 1977 off the Galápagos Islands. White smokers have also been found that erupt white calcium and barium.

Geological pseudocraters, Lake Myvatn, Iceland

THESE REMARKABLE PSEUDOCRATERS were formed in Lake Myvatn, Iceland, by lava-induced steam eruptions. The green vegetation-covered depressions are termed pseudocraters because they are not true volcanic craters formed naturally over a vent. Instead they were formed when, two thousand years ago, a fissure opened and lava infiltrated surrounding boggy wetlands. The water-lava combination set off a series of steam explosions that blasted the various pseudocraters up and out. When the lava drained the current Lake Myvatn filled the basin. Lake Myvatn is eurotropic, over saturated with phosphorus and nitrogen nutrient runoff, causing plants to overgrow. The result at Lake Myvatn are dense surrounding wetlands that have been turned into a nature reserve and are a haven for many bird species.

La Palma, Canary Islands

LA PALMA IS THE BIGGEST STRATO-VOLCANO in the Canary Islands. It rises 6,500m / 21, 300ft from the ocean floor and the underwater sections of the volcano are built up with layers of pillow lava. The top portion is made of alternating basalt lava and pyroclastic rocks. La Palma has had seven major historic eruptions and its summit is dominated by one of the largest (9 km / 5.6 miles wide x 1,500 m / 4,900ft deep) collapse craters on Earth. Some scientists have speculated that if La Palma had a big eruption a gigantic section of the volcano might collapse into the sea and cause a super-tsunami with waves reaching 900m / 300ft.

Ancient lava rock, Kartouche Bay, Anguilla, West Indies

THE WEST INDIES ARE A LONG CURVED VOLCANIC ISLAND ARC in the Caribbean Ocean that stretches from Puerto Rico to Venezuela. The arc was formed from the melting of the Atlantic Plate as it slips under the Caribbean Plate. The southern islands, called the Lesser Antilles, are still active, explosive strato-volcanoes, including Saint-Vincent's Soufrière, Guadeloupe's Soufrière, Montserrat's Soufrière Hills and Kick 'Em Jenny, a submarine volcano. The northern islands, which formed centuries ago by volcanic processes, are now extinct and covered with limestone. They appear idyllic now, but beaches like Kartouche Bay, Anguilla in the north, are littered with ancient volcanic rocks that echo their violent volcanic past.

Weird undersea lava formation

STRANGE THINGS HAPPEN WHEN BASALT LAVA ERUPTS under the sea. The cold temperatures and extreme water pressure cause the outer skin of molten lava flows to cool and solidify more quickly than at the surface, trapping explosive gasses inside. Hydrogen bubbles blast through the skin and deform the hardening flow. This causes the lava to solidify in weird shapes. Sometimes the skin cracks and hot lava oozes out cooling into unusual formations. Scientists are currently studying volcanic basalt rocks found on the deep ocean floor for traces of carbon dioxide to learn how volcanic eruptions affect global climate warming.

Divergent plate boundaries, Mid-Atlantic ridge

VOLCANIC MOUNTAIN RANGES CALLED RIDGES line sections of the Earth's sea floors and the Mid-Atlantic Ridge is one of these ranges formed along divergent (spreading apart) plate boundaries. This ridge slashes the floor of the Atlantic Ocean from North to South: in the North Atlantic the North American plate is pulling away from the Eurasian Plate; in the South Atlantic the South American Plate is moving away from the African Plate. Along the widening seam erupting lava has built giant volcanoes over millions of years that have broken the ocean's surface forming emergent islands including Iceland, Jan Mayan, and Ascension.

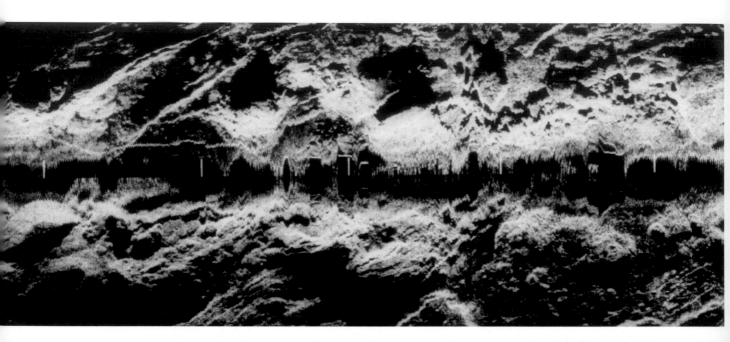

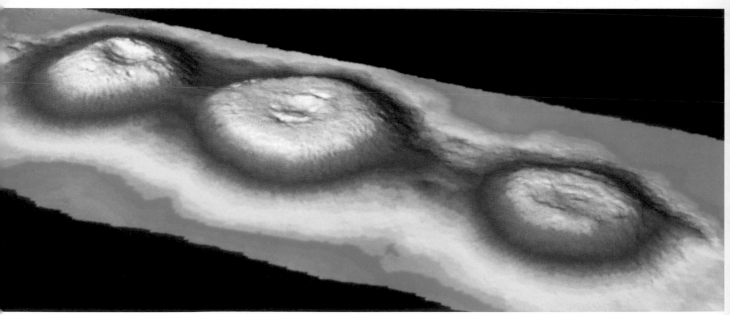

Row of submarine volcanoes, The Three Wise Men

THE FLOOR OF THE PACIFIC OCEAN has more seamounts, or undersea volcanoes, than any other ocean. Undersea volcanoes, as other volcanoes, form from molten lava produced where the edges of Earth's plates bump into, slide under or grate against each other. A row of three submarine volcanoes, called The Three Wise Men, grew along a spreading mid-ocean ridge called the East Pacific Rise. The Three Wise Men are part of a seamount field called Rano Rahi in the Pacific Ocean. The Rano Rahi seamount field is known for its numerous chains of submerged volcanoes. The Three Wise Men are between 1,000 and 2,000m / 3,280 and 6,560ft tall. In this image, green indicates deep areas and red and white more shallow.

Black sand beach, Dyrholaey, Iceland

WHEN FLUID BASALT LAVA that is very hot (up to 1,200°C / 2,192°F) flows directly into the cold ocean it explodes and shatters into sharp tiny pieces. The glassy lava shards are carried by ocean currents and deposited in pockets along the shore as coarse black sand beaches. If the sand is carried a long way before being deposited, the action of waves and currents makes the sand grains rounder and smoother. For many years professionals theorized that black lava sand beaches formed over long periods of time by the slow erosion of solid lava, but in the 1980's as hot lava poured into the ocean at Kamoamoa, Hawaii, heavy wave action blasted it into sand grains, creating a stunning black sand beach in one weekend.

Hanauma Bay, Hawaii, USA

FOUR MILLION YEARS AGO THE NORTHERN ISLAND OF OAHU, Hawaii rose above the Pacific Ocean as Waianae volcano, that was fed by a hot spot plume of magma melting through the Pacific Plate. Hanauma Bay, today a popular reef snorkeling area, was blasted open by a hydromagmatic explosion when an active vent poured magma underwater. As the plate slid northwest it dragged Oahu and its volcano off the hot spot and the volcano went extinct. The cone of Hanuama wore away by wave action and the sea flooded in. Beautiful coral reefs then grew in the cone left at Hanauma forming a protected habitat for a multitude of sea creatures.

Surtsey bursts from the sea

SURTSEY, A CLASSIC SUBMARINE VOLCANO THAT FORMED A NEW ISLAND, grew out of a volcanic fissure along the Mid-Atlantic Ridge. The eruption was first sighted on November 14, 1963 as a column of black smoke rising from the sea 2 to 3 km- / 1.2 to 1.8 miles high. Ships investigating the scene found an explosive eruption of ocean water and black ash shooting skyward and in less than a week a 45m / 147ft high island had formed. Submerged eruptions of pillow lava piled up 130m / 426ft from the sea floor to the ocean's surface and by early 1964 so much lava had been erupted and cooled to stone that the island was permanent. Surtsey is named after the Norse god of fire. Today 50 species of plants and several bird species, including seagulls, swans, geese and puffins, call Surtsey home.

Aa lava flow meets the sea, Galápagos Islands, Ecuador

THE GALÁPAGOS ISLANDS, A STRING OF BASALT SHIELD VOLCANOES, about 3-5 million years old, are young by geological standards. They sit on a hot spot where magma from inside Earth burns through the crust feeding six active
volcanoes. Over 49 eruptions have occurred here in the past 200 years. Many of these were fissure eruptions, that
disgorged huge amounts of fluid lava forming sheets and plateaus, rather than single vent eruptions. These forceful high-volume flows advance tens of kilometers to pour into the sea. Both smooth pahoehoe and thicker, block-
ier aa flows, like the one in this image, erupt from Galápagos' volcanoes. The clinkery surface of aa seals in a
molten core that rolls forward with a conveyor belt motion.

Pangea, Godwanaland and Laurasia

IN THE EARLY 1900s Alfred Wegener proposed a theory, based on the global distribution of fossils, that stated; over 500 million years ago Earth had one super continent, surrounded by ocean, called Pangea. Pangea cracked apart creating Godwanaland in the southern hemisphere and Laurasia, which drifted to the north. These two chunks continued to break, like matching puzzle pieces, and drift apart, creating the continents we know today. In the 1960s, when sea floor spreading was scientifically documented, Wegener's theory was vindicated. Today the science of Plate Tectonics explains how a dozen drifting crustal plates, bump, crash, slide, grate, dip and push against one another freeing molten lava from inside Earth to erupt at these volatile seams and give birth to Earth's volcanoes.

WITH THE ADVENT OF THE SPACE AGE, scientists have unearthed a mountain of evidence that shows that our solar system is overflowing with volcanic activity. In 1969, Apollo 11 astronauts took the first steps on our nearest neighbour, the Moon, atop a time-worn lava flow in the Sea of Tranquility and over the following three years, astronauts brought dozens of volcanic rocks back from the Moon for study. Although our Moon currently has no known active volcanism, it has features that exhibit clear evidence of past volcanism, such as vast basalt flows, called maria (literally 'seas'), as well as collapsed lava tubes, called rilles and volcanic cones, or domes. Space missions to Mars have also sent us back thrilling images of several mighty volcanic edifices there. These include Olympus Mons, a giant shield volcano measuring 550 km / 340 miles across and 24 km / 15 miles high, that is the largest known volcano in our solar system. The Magellan space probe to Venus brought back amazing three dimensional images from there of some of the strangest volcanoes and longest lava flows channels in our Solar System. And though we have not absolutely confirmed whether Venus is still volcanically active, 90 per cent of its surface is known to be covered in basalt. Venus, then, exhibits possible rocky volcanism similar to that found on Earth. Elsewhere in our solar system, Mariner 10 spacecraft images of Mercury have revealed what appear to be smooth volcanic plains on its surface and Voyager's images of Neptune's moon Triton, show signs of volatile volcanism probably caused

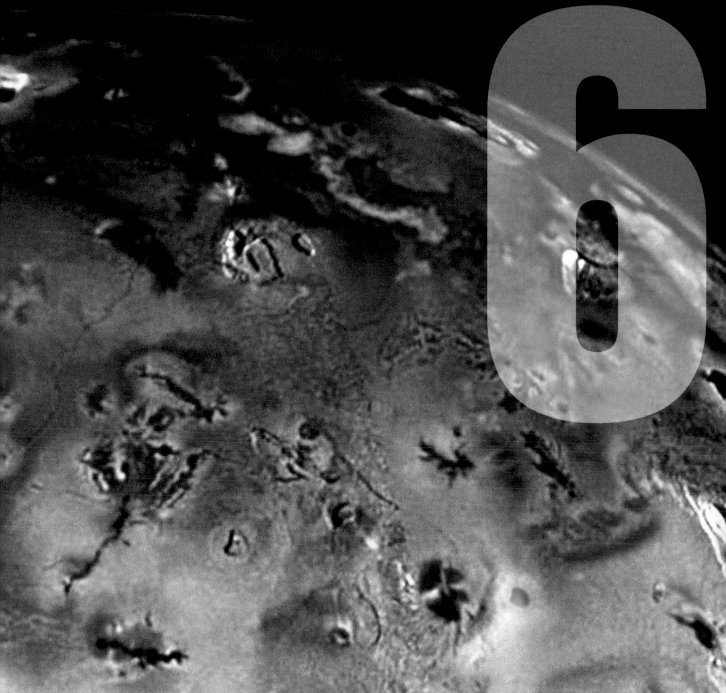

6

from nitrogen and methane gasses. Jupiter's Moon Io is one of the most volcanically active worlds in the solar system besides Earth and in February 2001, the largest recorded volcanic eruption in the solar system occurred here. Astronomers believe Io has sulfur eruptions, that are the result of its interior being heated by the massive gravitational tug of the giant Jupiter. Io's 1500°C /2730°F lava is some of the hottest molten rock ever recorded.

SPACE VOLCANOES

Jupiter's moon Europa has ridges that could be the frozen remnants of cryovolcanic activity, when water or water ice erupted on the Europan surface and even comets are pocked with features that appear to be volcanic-like vents. These vents are the heart of subsurface eruptions, where ices sublimate (transform from ice directly to a gas) into long streaming jets. The study of volcanoes on other planets and moons within our solar system, in conjunction with the study of those on Earth, helps us to unravel the mystery of how Earth and the other planets and moons in our Solar System formed.

Caloris Basin and rilles, Mercury

MERCURY'S CALORIS BASIN, one of the largest impact craters in the Solar System – 1,300km / 807 miles in diameter – is ringed by mountains 2km / 1.2 miles high. It was formed by an impact of a magnitude great enough to trigger volcanic events – the floor of Caloris Basin is engraved with cracks, fractures and serpentine ridges called sinuous rilles, believed to be collapsed lava tubes or lava channels. Smooth plains inside the basin are stippled with what look like lava-flooded pit craters, and these appear to have been caused by volcanic collapse rather than by an impact. The broad, smooth plains inside the basin also appear to have been created by lava floods.

Lava flows, Mercury △

SCIENTISTS WONDERED FOR YEARS if Mercury, like the Earth, Mars, Venus and the Moon, had a history of volcanism. The Mariner 10 spacecraft gave them some answers. As the craft sailed by Mercury it snapped images that showed craters and smooth plains that scientists now interpret as over-spilling lava flows. If this is the case, Mercury has had active volcanism in the recent past. Scientists suggest that jetting lava fountains, propelled by volatile gasses, may have occurred on the surface which may, at one time early in its history, have been covered by a molten ocean. The Mariner 10 mission gathered about 7,000 images of Mercury, Venus, Earth and the Moon.

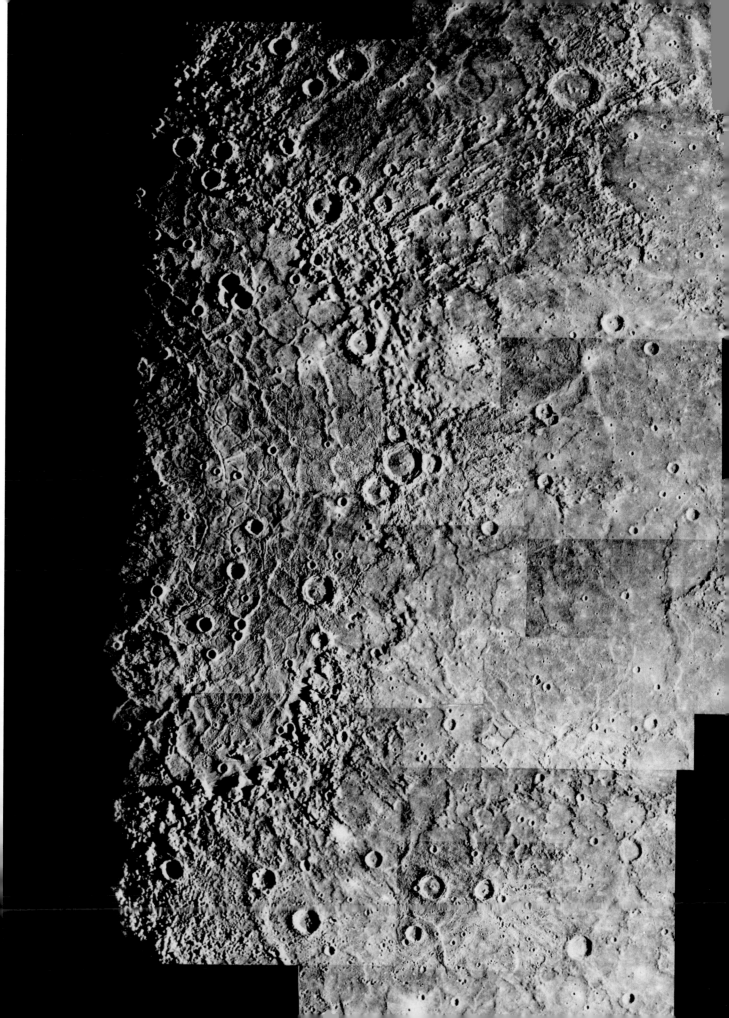

Highest mountain and longest lava flows, Venus

VENUS'S SURFACE IS SPECKLED with mountains, plains, craters and lava channels. One of the longest lava-flow channels in the Solar System – measuring over 7,000km / 4,350 miles long and making it longer than the River Nile – dissects the planet's face, and is believed to have been formed when lava traversed the vast plains. Sections of the main lava channel are now covered by newer lava flows. The bright spot, Maxwell Montes, in the upper-left-hand section of this image, is the tallest mountain on Venus at 11km / 6.8 miles high.

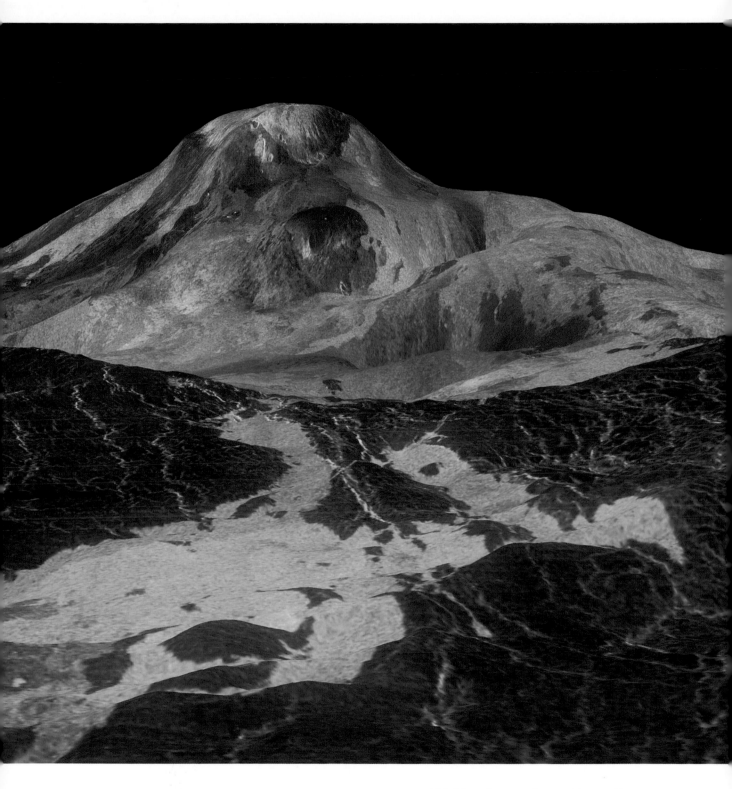

Shield volcano, Maat Mons, Venus

VENUS, EARTH'S 'SISTER' PLANET, has a similar size and orbit but, unlike Earth, the surface temperature is 470°C / 878°F, its atmosphere is carbon dioxide, and its clouds are composed of sulphuric acid. Planetary geologists believe that huge Plinian eruptions injected Venus's atmosphere with sulphur dioxide and methane and in fact Venus has more than 100,000 volcanoes, more than any other planet in the Solar System. Most are shield volcanoes like Maat Mons (the tallest at 8km / 4.9 miles high), which has a linear string of collapse craters dotting its south-west flank, indicating a rift zone. The Magellan space probe found evidence of recent ash eruptions near the summit, as well as evidence that Venus's other volcanoes are still active.

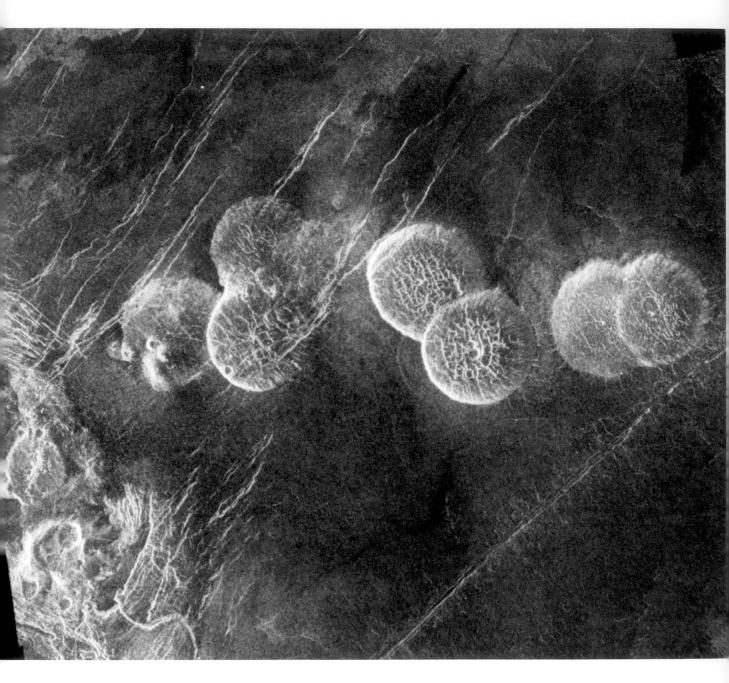

Pancake domes, Alpha Regio, Venus

IN ADDITION TO ITS SHIELD VOLCANOES, Venus has some strange volcanoes that may erupt thick quartz or granite. Seven bulging, pancake-like structures form a cluster in Venus's Alpha Regio region. They appear to be the result of eruptions of thick, paste-like lava that sat over vents and ballooned up, giving them their characteristic domed shapes. Each round hill is about 25km / 15 miles wide and 750m / 2,395ft high, and there are cracks and fractures visible on the outer skin of the domes. The Magellan space probe was equipped with a special sonar-like imaging system (radar altimetry) which gives us a three-dimensional view of the planet's surface, such as in the image above.

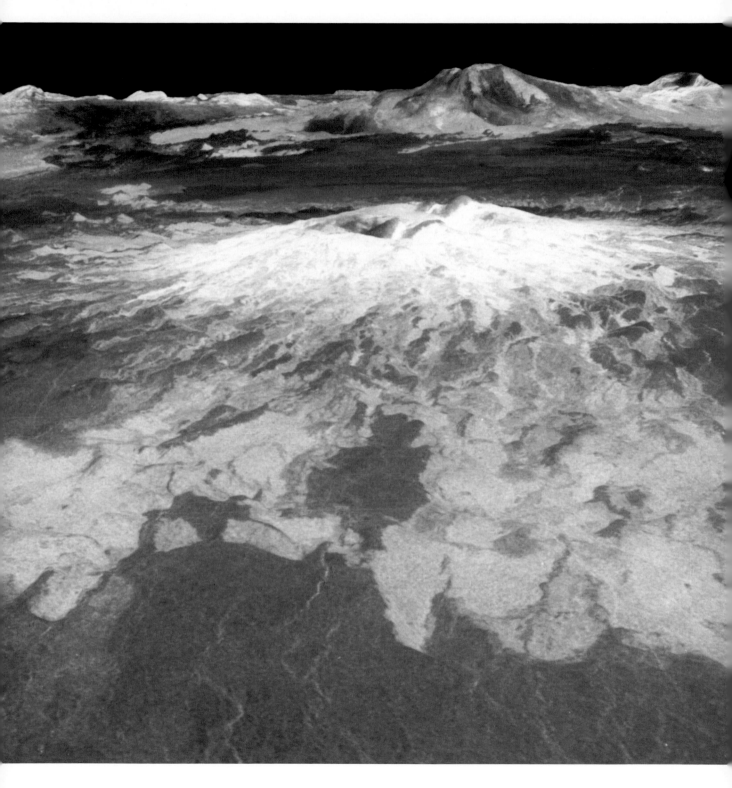

Hawaiian-type volcano, Sapas Mons, Venus

VENUS IS HOME TO MORE THAN 150 GIANT, broad, shield volcanoes similar to Hawaiian-type volcanoes on Earth, but much bigger. Sapas Mons – 400km / 248.5 miles in diameter and 1.5km / 0.9 miles high – is one such huge, gently sloping shield volcano in Alta Regio with a twin, flat topped summit. Hundreds of fluid, flank lava flows zig-zag down its broad sides. Due to their highly scattered distribution, most shield volcanoes on Venus probably sit above individual mantle hot spots, unlike the vast majority of Earth's volcanoes that form in lines along tectonic plate rifts.

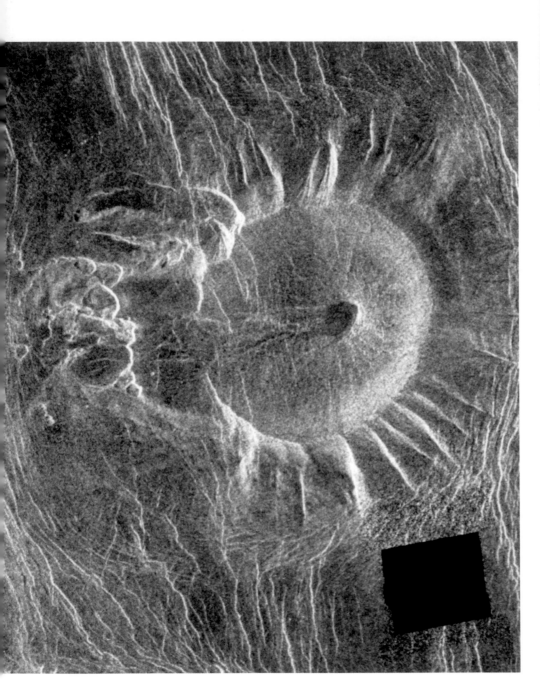
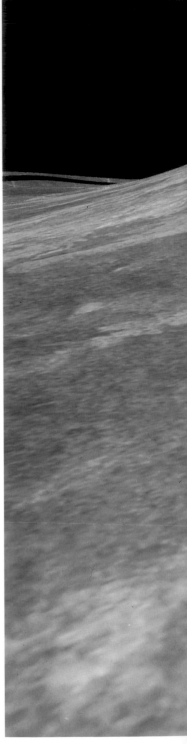

Tick volcano, Eistla, Venus

THE TICK, NAMED BECAUSE OF ITS SHAPE, is a concave volcanic feature – about 35km / 22 miles wide across its abdomen – in Venus's Eistla Regio. A crater or pit is visible in its centre, the 'legs' are alternating radial ridges and fluted valleys, and the 'head' is a cluster of pit craters. The overlapping layers seen in the left-hand side of the image are probably caused by successive lava flows. There are half a dozen similar tick-like features on Venus, and no one knows exactly how they were formed. Scientists think that they could be collapsed and eroded pancake domes.

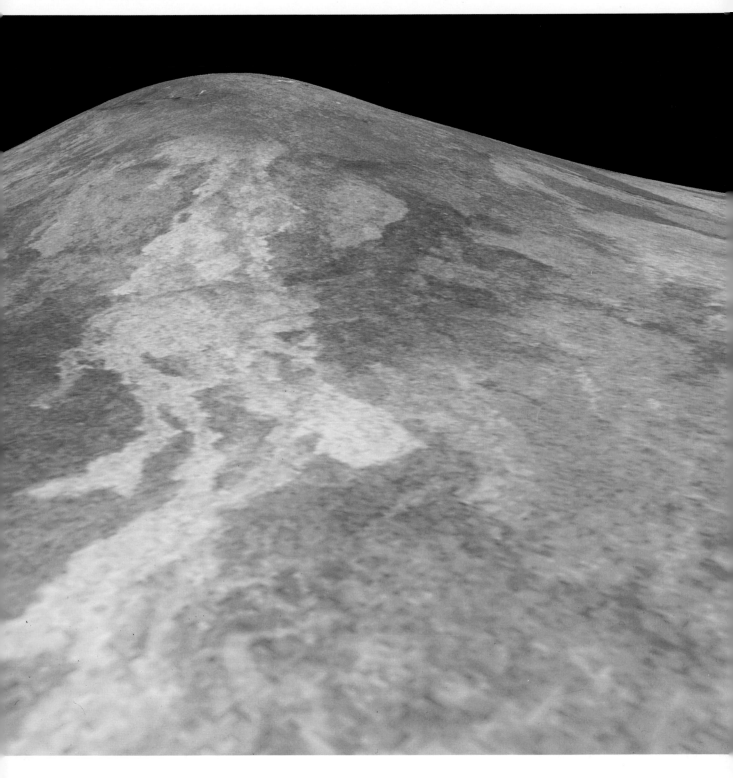

Sif Mons, Venus

THE MAGELLAN SPACE PROBE took wonderful colour-enhanced radar images of Sif Mons volcano on Venus. Fluid lava flows, coloured light yellow, hundreds of kilometers long, can be seen flowing down the volcano's flank towards the foreground. Sif Mons, over 300km / 186 miles wide and 2km / 1.2 miles high, towers over the region called Western Eistla Regio, near the equator and is believed to be a hot-spot volcano like Kilauea, in Hawaii. Scientists speculate that, because most of Venus's shield volcanoes sit on hot spots in cleaved rift zones, long tears and cracks in the crust may be required in order for volcanoes to form on Venus.

THE MOON'S MAJOR VOLCANIC ACTIVITY took place about 4 billion years ago. It is believed a space object slammed into the Moon, punched out the giant 1,123km- / 698 mile-diameter crater, Mare Imbrium (Latin for oceans and rain showers), cracking the Moon's spherical lithosphere like an egg shell, and releasing lava flows that filled the crater. Although there are immense basalt lava flows blanketing the Moon's surface, there are no big volcanoes. Scientists believe that the Moon's low gravity pull, one-sixth of Earth's, means that eruptions are blown in a widespread area and don't build up into a cone shape. Astronauts on the Apollo 17 mission brought 3.7-billion-year-old Mare basalt lava back to Earth.

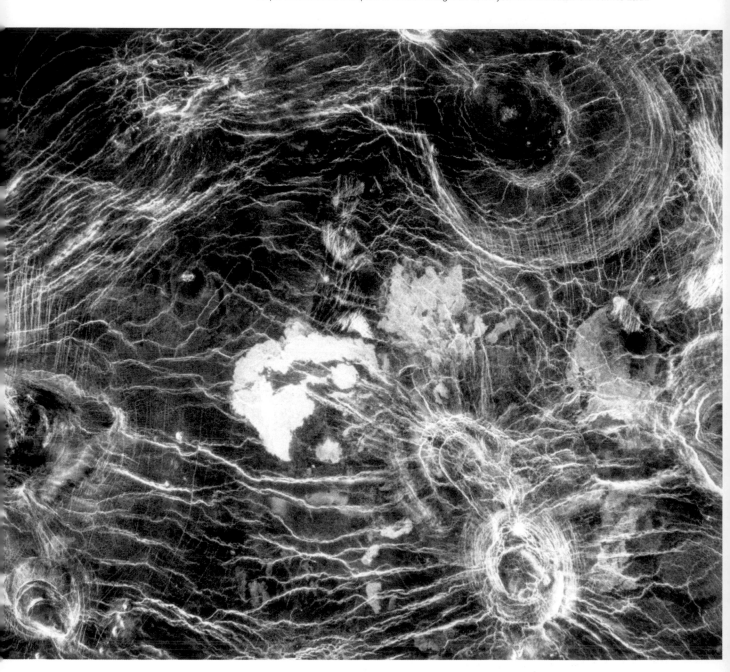

Arachnoid volcanoes, Venus

UNTIL SPACE PROBES ALLOWED us to see below the clouds of Venus, we could only make wild guesses as to what might be found on its surface. These strange volcanoes, about 100km / 62 miles in diameter, have a spidery cobweb appearance from above, hence their name. Each spider has a circular, volcanic, crater-like edifice at its centre, with radiating, fractured lines fanning out and away like a wispy web. No one knows the volcanic processes of arachnoid volcanoes, and they have been found nowhere else in the Solar System. Venus has four groups (and there are about 30) of arachnoids that are found in regions of compressed and faulted plains.

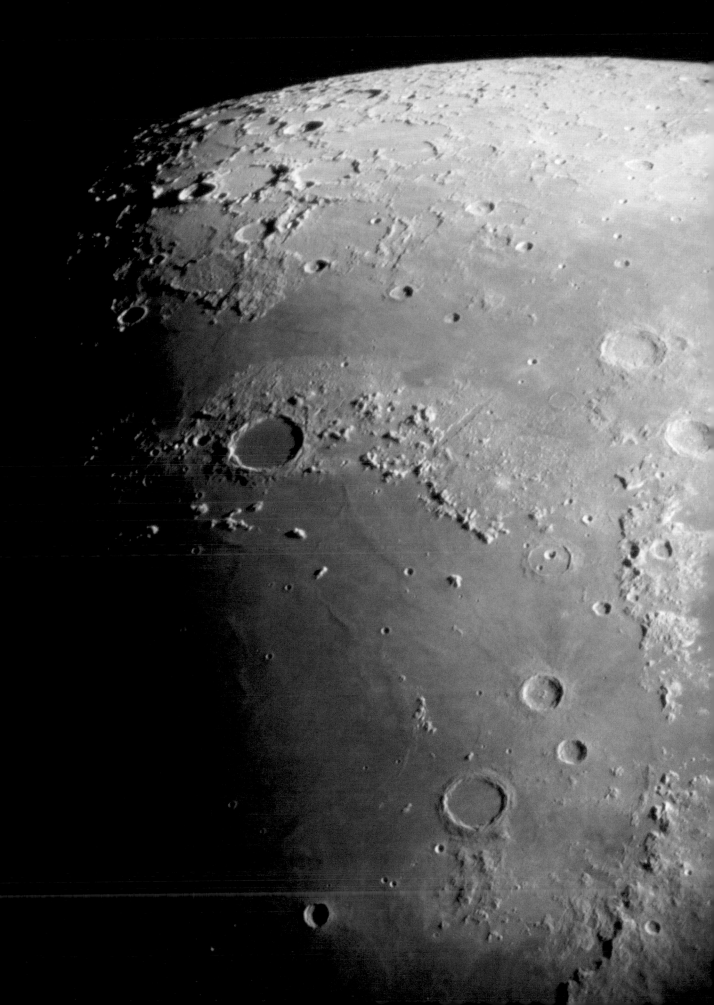

Sinous rilles, the Moon

FOR DECADES, ASTRONOMERS LOOKED THROUGH telescopes at the Moon and debated whether the winding valleys - up to 300km / 186 miles long and 1,300m / 808ft deep – were rivers. The truth is that the Moon is dry as a bone, but has a volcanic past. The surface is creased with hundreds of volcanic land forms that look like snaking channels, called sinuous rilles, and many originate from craters. Astrogeologists believe they are either collapsed lava tubes, or lava channels caused by hot flows that thermally ate away the Moon's surface. The Moon is Earth's only natural satellite.

Volcanic soil, lunar surface

THE WORLD HELD ITS BREATH in 1969 when astronauts made the first boot prints in the soft lunar soil. Meteorites have continuously bombarded the Moon and have pulverized lava flows into grainy soil. Hundreds of kilograms of it brought back to Earth have been shown to be igneous, being made from cooled lava containing basalt, and green and orange glass particles. This image shows ancient volcanic basalt-titanium lavas in blue, lavas containing small amounts of titanium in orange and scattered explosive pyroclastic debris in deep purple. Instrumentation left on the Moon records about 3,000 moonquakes annually. Scientists studying lunar soil samples have found no evidence of life.

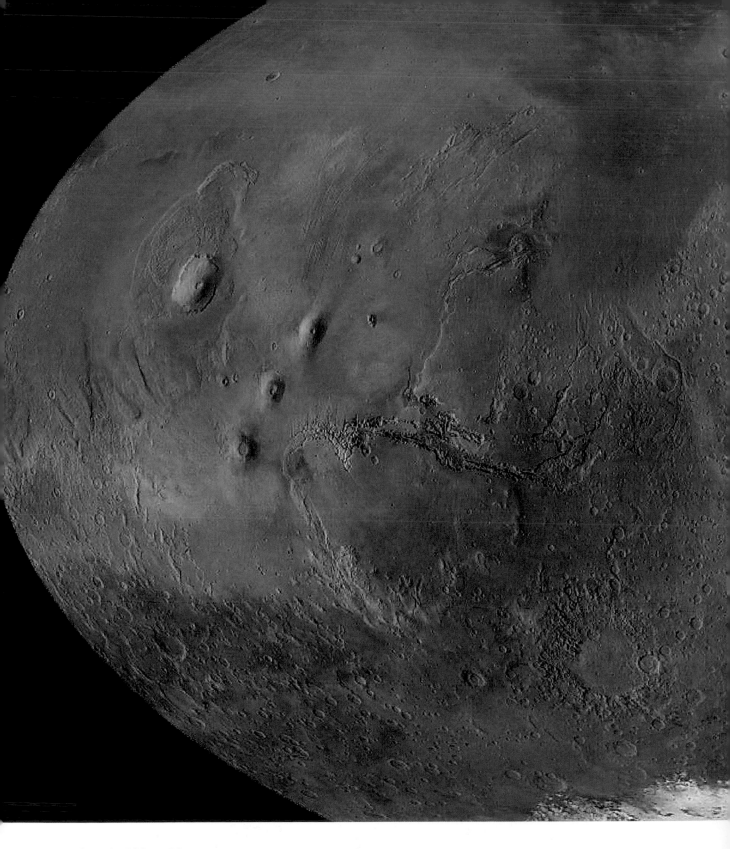

Tharsis Ridge, Mars

DESPITE MARS'S SMALL SIZE, it is a highly volcanic planet. Early in its history volcanoes broke through surface faults, lava deluged the surface and a huge balloon of hot, rising, internal planet material gave birth to the Tharsis Ridge. Three major shield volcanoes, Ascraeus Mons, Pavonis Mons and Arsia Mons, form a line on Tharsis Ridge, covering an area about twice the size of the USA, and the entire Ridge actually swells up 10km / 6.2 miles from the surface. Mars is pocked with meteor craters and, occasionally, the planet's rocks are jettisoned into space by these impacts and travel as far as Earth. For the last several hundred million years Mars has not been volcanically active. This pole-to-pole Mars image was digitally constructed from 200 million laser altimeter measurements.

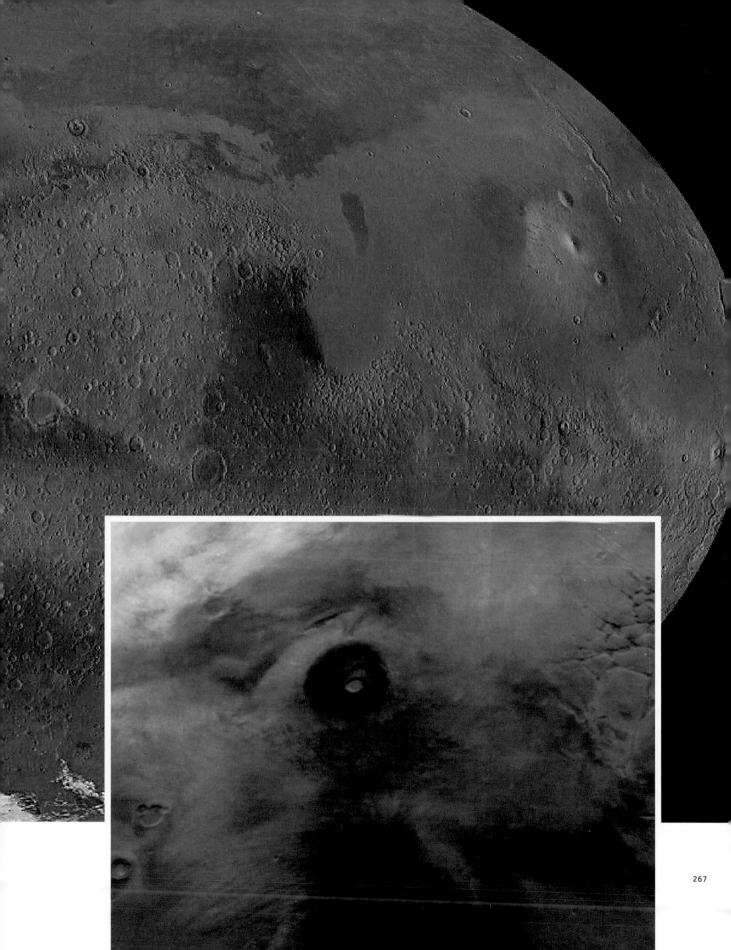

Largest volcano in the Solar System, Olympus Mons, Mars

THE UNBELIEVABLY VAST OLYMPUS MONS, is on tiny Mars. At 27km / 16.7 miles high and 550km / 342 miles wide, it's three times taller than Mt. Everest, and is about the same width as the state of Arizona in the USA. Olympus Mons may still be volcanically active and, like Mauna Loa in Hawaii, is believed to have formed over a volcanic hot spot. Unlike Earth, Mars has no tectonic plates, so the massive amounts of fluid lava continually erupting in stationary areas bulge up into huge, gently sloping volcanoes. This volcano is so large that it has often peeked above the frequent sand storms, and it was even visible to nineteenth-century astronomers.

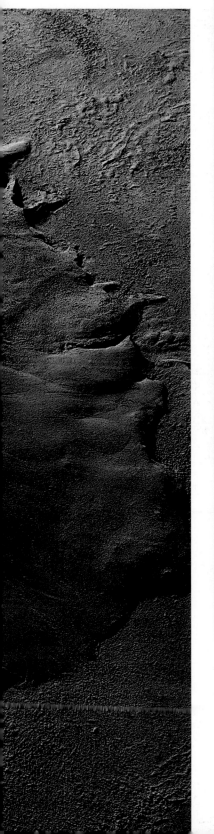

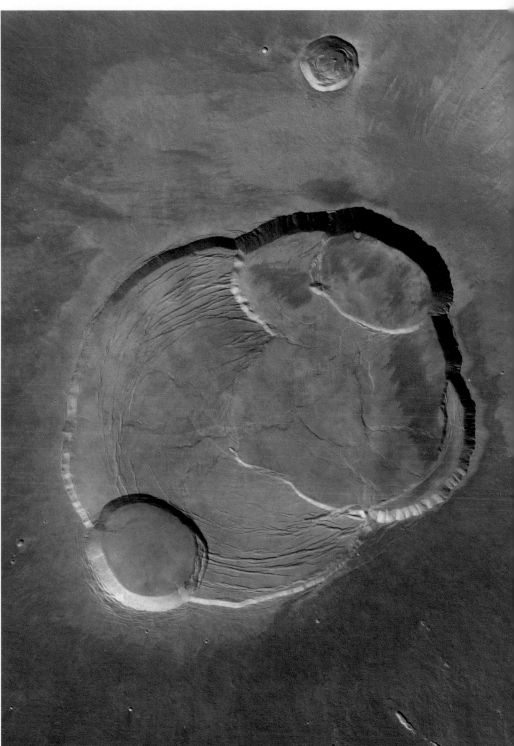

Olympus Mons caldera, Mars

THE CALDERA OF OLYMPUS MONS IS HUGE: 85km / 52.8 miles wide and 3km / 1.8 miles deep, with six overlapping pit craters inside. When the last eruption ceased and the deep magma chamber emptied its contents, the massive caldera collapsed into the monstrous void. Scientists estimate that the magma chamber was over 60km / 37.2 miles wide dwarfing those found on Earth. Additional collapses happened at the outer edge of the caldera giving it a jagged look. The original collapse also caused deep fractures radiating from the summit caldera. Olympus Mons caldera is similar to collapse calderas found on Earth's shield volcanoes.

Basalt lava boulders, Mars

BOULDERS AND ROCKS FOUND ON MARS are weirdly like those on Earth. When the Sojourner Mars Rover used a high-tech X-Ray Spectrometer to examine a boulder, it was shown to be similar to the most ubiquitous rocks found on our planet, terrestrial basalts. Smooth, fine-grained, silica-poor basalt is solidified from fluid lava erupted from vents and fissures, and NASA reports that basalt is the most common rock in the inner Solar System. NASA even named one large basalt boulder on Mars 'Yogi' after the baseball legend Yogi Berra because it has the same composition as rocks found directly under the Yogi Berra Stadium in Montclair, New Jersey, USA. The only difference is that Mars's rocks are about ten times older than terrestrial basalts.

Blueberries, volcanic lapilli, Mars

THESE PELLET-SIZED BLUE SPHERES, dubbed by minarologists 'blueberries', made of iron-rich hematite, were found at the Mars Rover Opportunity's landing site, 'Shoemaker's Patio', amidst small bits of volcanic debris. On Earth, hematite forms only in wet conditions and scientists think that these small blue balls formed during the accretion of soggy, mineral sediments, building up in layers on their surface. If the blueberry balls are indeed concretions, this would prove that Mars once had water. The dark sand also found at the site, also seen here, is volcanic lapilli, tiny black particles of ash ejected from volcanoes.

271

◁ Pele volcano, Io ▷

JUPITER'S CLOSEST MOON, Io, the first active volcanic body discovered in our Solar System, was a total surprise. One of the largest visible features on Io is Pele volcano, named after the Hawaiian volcano goddess. A massive 1,400km-/ 869.9 mile-diameter halo of recently erupted, red volcanic sulphur debris encircles Pele, and black blemishes surrounding it are evidence of recent volcanic activity. The volcano has been known to launch eruptions more than 400km / 248.5 miles above the surface in a dome-shaped blast, and Pele's central vent is believed to hold a super-hot lava lake. The Galileo spacecraft snapped images of glowing red cracks exposing hot rifts of lava at the summit.

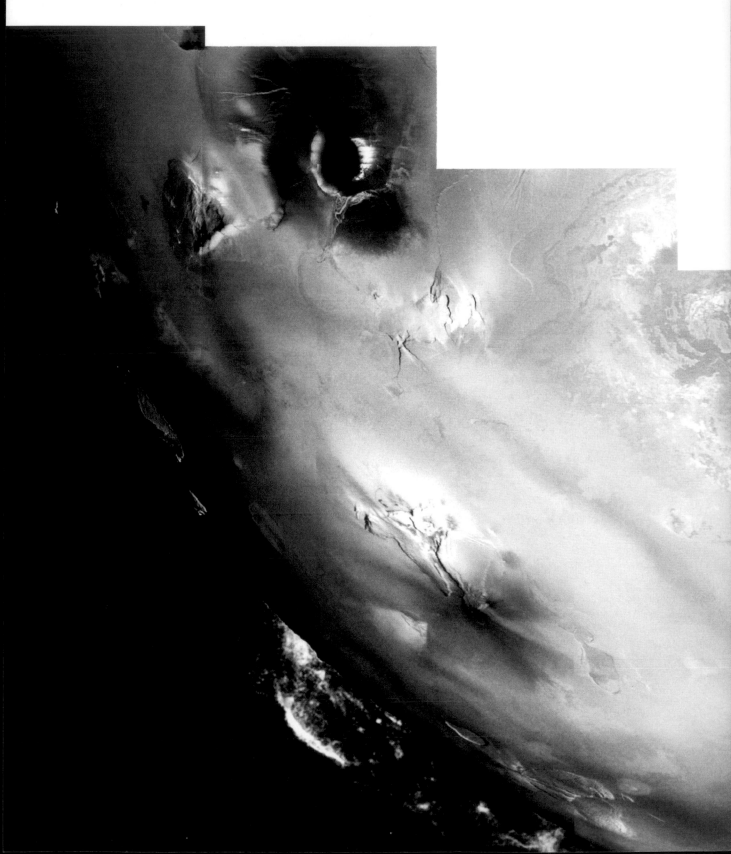

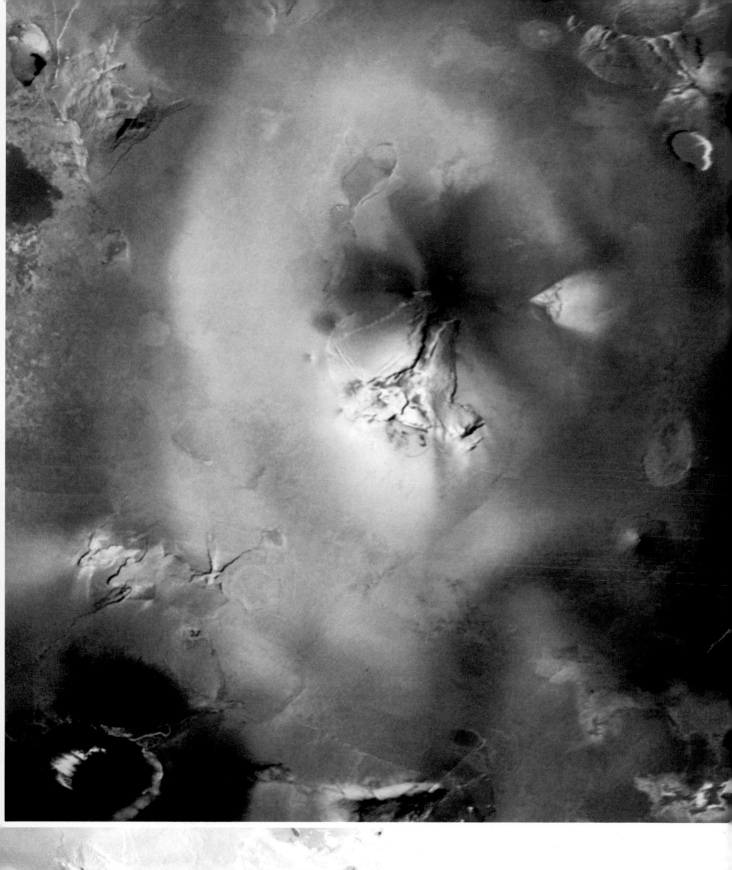

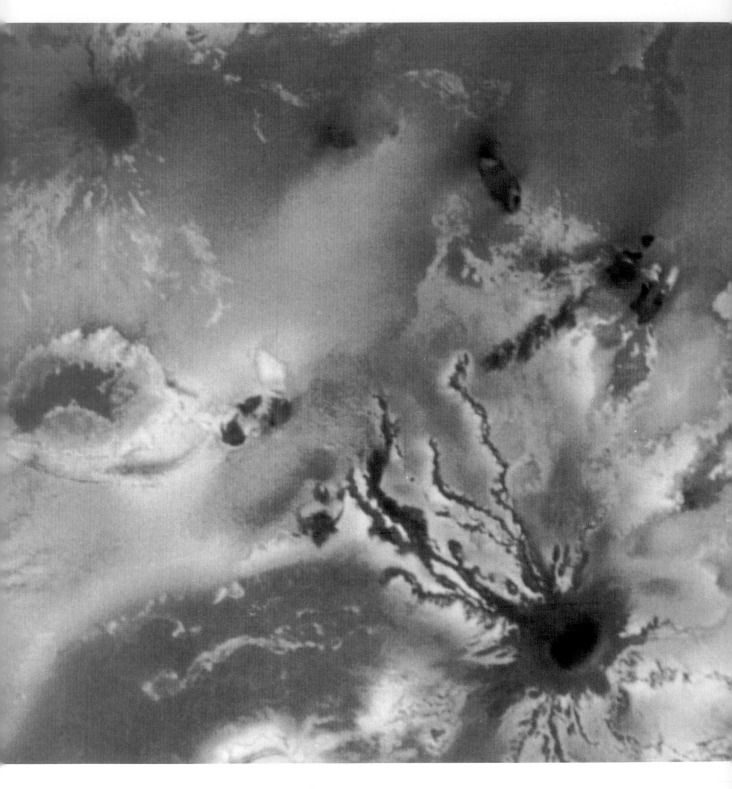

Longest active lava flows in Solar System, Io

THE LONGEST ACTIVE LAVA FLOW – at 250 km / 155.3 miles – discovered in the Solar System snakes across the hot volcanic terrain of Amirani on the Jovian moon Io. New lava continually feeds the flow and like Earthly lava, Io's flows travel great distances from the spewing vents in tubes under the hardened surface. The longest lava flows on Earth, found at Kilauea Volcano in Hawaii, are miniscule when compared with Io's. Spacecraft images have revealed sulfur gasses leaking to the surface across Amirani's plains and what appear to be massive curtains of fire. By studying these monstrous lava flows and surface features, scientists can get clues as to how the Earth may have formed.

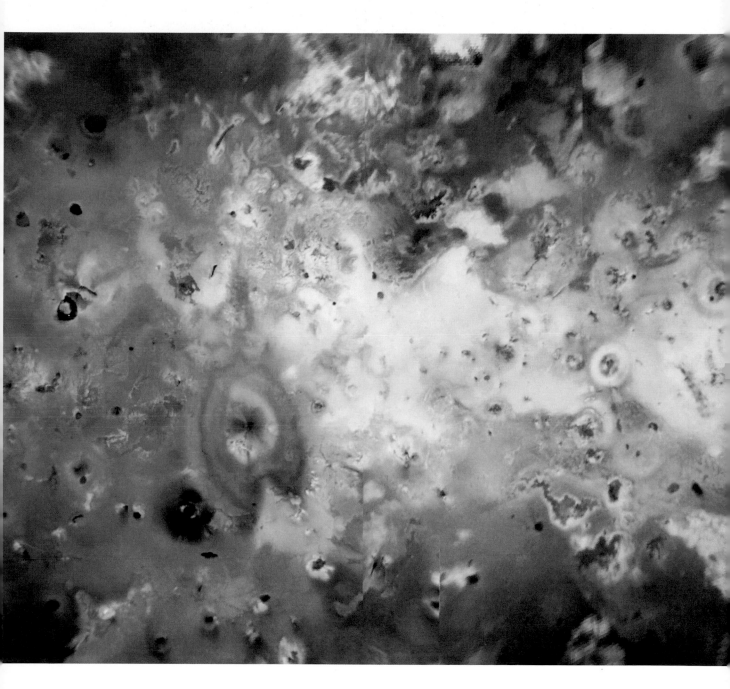

White sulphur snow deposits, Io

SULPHUR IS ONE OF THE MOST common elements in the Solar System and Io's volcanically turbulent surface is saturated with it. Red and black volcanic debris rings have soft white halos – deposits of sulfur-dioxide, erupted as hot vapor and frozen solid as sulfur frost and snow. The dark ring in the upper left of this image is an active lava lake in a volcano named Loki Patera. Loki is more powerful than all of Earth's volcanoes merged together. Io is so violently volcanic because it is continually deformed and contorted by the great gravitational pull of giant Jupiter and its neighboring moons.

Volcanic plumes and geysers, Io

GIANT BALLOONING VOLCANIC VENTS on Io belch plumes of gas in eruptions that work like geysers on Earth. Instead of squirting water, Io's vents blast hot sulfur-dioxide gas more than 500km / 310.6 miles into Space. As the hot gasses and particles burst upward they fast-freeze into sulfur dioxide snowflakes as captured here during the Voyager space mission. They look like shimmering translucent domes rising from Io's limb. Io is one of four Galilean moons discovered in 1610 by the great astronomer Galileo Galilei. Seeing the moons orbit Jupiter helped him move forward his theory for a Copernican, sun-centered, solar system – one in which everything did not revolve around Earth as was then believed.

Pyroclastic deposits and constant plume, Prometheus, Io

PROMETHEUS VOLCANO has such a huge and constant eruption plume it has been called Io's Old Faithful. The umbrella-shaped plume has remained the same size (85km / 52.8 miles tall) for 20 years and appears to be erupting from a sizzling dimpled vent. Prometheus and Io's 70 other active volcanoes are believed to be responsible for covering large regions of Io's surface with enormous red deposits, which astronomers think are sulfur-dioxide-laced, pyroclastic matter. Because of all this volcanic activity Io's surface is one of the most colorful in objects in the Solar System. In this NASA image two plumes are visible: one at the limb, from Pillan Patera Volcano, and one in the center (grey with a black shadow), from Prometheus volcano.

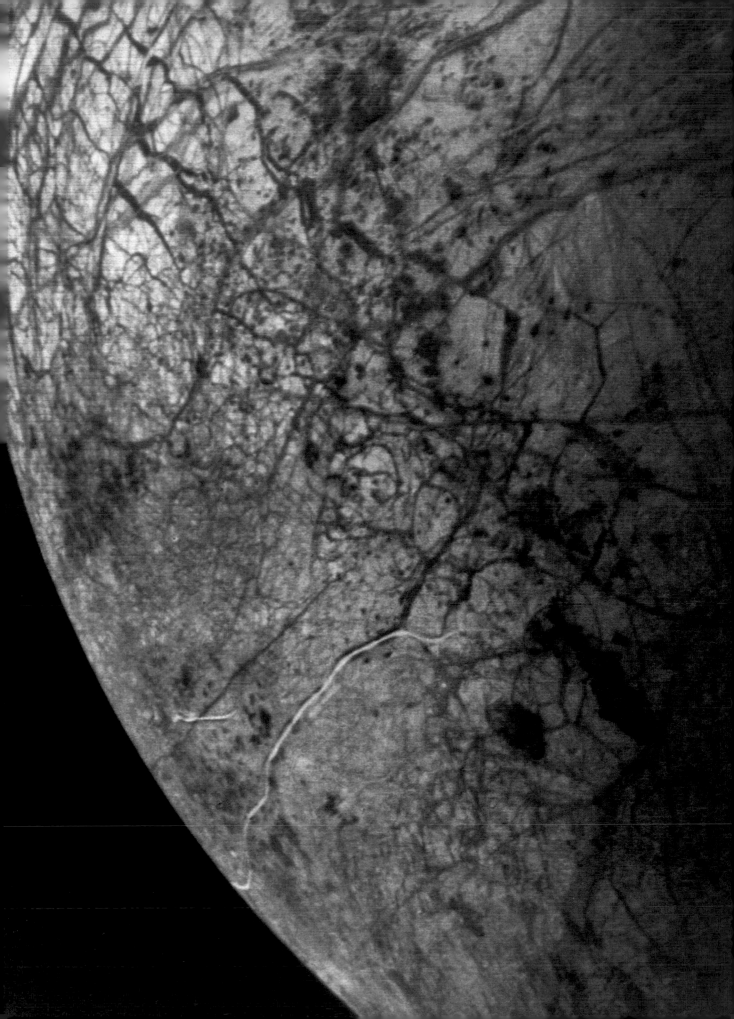

Cryovolcanic ice moon, Europa

JUPITER'S MOON EUROPA travels so far from the warmth of the Sun that it is a bitterly cold (-163°C / -261°F) arctic world. It is one of the smoothest objects in the Solar System and its icy surface has few, if any, craters. Thin red cracks, formed by the tremendous gravitational pull of Jupiter, stretch thousands of kilometers across the frigid moon's face. Europa exhibits cryovolcanic activity, meaning it erupts volatile liquids like water and ammonia instead of hot lava. When this icy liquid and vapor sludge, called cyromagma, arrives on the harsh, cold surface, it freezes instantaneously. Blue ice under the crinkled surface is believed to be a frozen ocean of water 100km / 62 miles thick. Europa is thought to have a core of silicate rock.

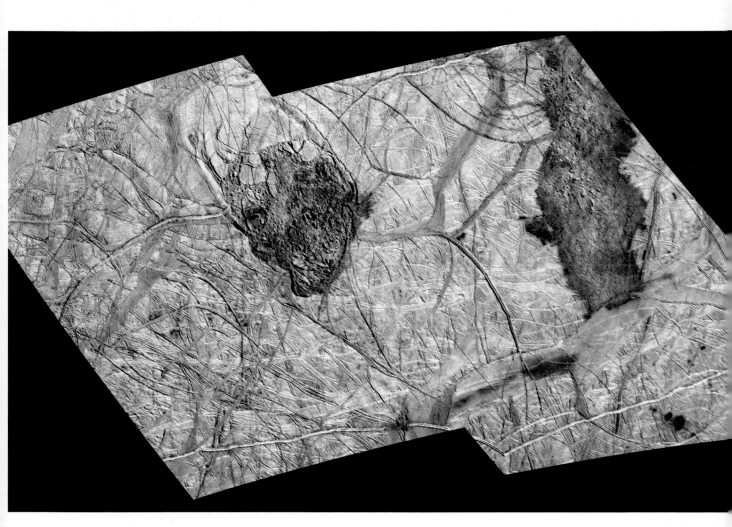

Thera and Thrace regions, Europa

TWO BROWNISH-RED REGIONS on icy Europa stand out against the moon's smooth icy face. Theories abound about how these features formed. It's possible that a warm, deep ocean under the icy outer shell, heated by Europa's insides, melted through the surface creating these dark regions, which scientists have labeled Thera (left) and Thrace (right). Rafts and slabs of ice inside these dark regions appear to be fractured. They stick up like ice-thaw seen in Earth's polar seas during the spring melt. Because Thera has a scalloped edge it has led some to speculate the spot was caused by collapse. For years scientists have ventured that if Europa has a subterranean sea it could harbor extraterrestrial life.

Early volcanism, Dione

THE CRATERED ICY DISC of Saturn's moon Dione was volcanic in its early history: cryovolcanism was responsible
for creating Dione's fragile outer shell. Cooling and heating of the moon's interior – made chiefly of water ice with
a hard rocky core – further compressed and fractured the surface. Brilliant, delicate streaks etched atop Dione's
dark surface are slumped cliffs of ice formed from ice volcanoes and massive tectonic fracturing. Dione's frozen
terrain is bespattered with impact craters wider than 35 km / 21.7 miles in diameter, so large that astrogeologists
speculate they could have spun the moon on it axis.

▽ Volcanic vents, Halley's Comet and Comet Wild jet release ◁

COMETS HAVE MESMERIZED mankind for centuries. As the frozen nuclei of comets like Halley and Wild approach the torrid sun they begin to defrost by sublimation, meaning they change directly from solid ice into gas without a liquid phase. Gasses trapped beneath the comet's dynamic surface are jettisoned from volcanic-like vents and escape into space as elongated streams of gas and dust millions of kilometers long. The fleeing dust and gas form a comet's signature tail, which always points away from the sun. Normally we see the comet's dust tail reflecting sunlight as it blazes across our night sky. Comets are dirty snowballs created about the same time as our planets. Modern scientists have always suspected comets might be sources of water and other components for life in the early Solar System.

GLOSSARY

aa Hawaiian word describing clinkery, basalt lava flow with angular chunks.

andesite Volcanic rock characteristically medium to light in color and containing 54 to 62 per cent silica.

ash Fine-grained fragments (< 2mm / 0.1in) of volcanic ejecta, mainly with a high glass content.

basalt Dark volcanic rocks or lava that contain 45 to 54 per cent silica.

bench The unstable, newly-formed, leading edge of a lava delta.

blast An explosive eruption.

block A squarish, angular, hunk of lava tossed out during an eruption.

bomb Semi-soft incandescent lava lumps (> 64mm / 2.5in) that take on a spherical stream-lined shape with tapered ends, when ejected.

caldera A big (> 1.6 km / 10 mile-wide) bowl-shaped depression formed by collapse when the underground magma chamber empties, or by a big explosive eruption.

cinders Irregular shaped bits of lava and lava rock, 64mm / 2.5in to 30mm / 11.8in wide). Sometimes called scoria.

cinder cone A hill with a bowl-shaped depression on top, built around a volcanic vent, composed of cinders tossed from the vent.

continental drift This theory states that continents move toward or pull away from each other horizontally.

complex volcano A volcano that has more than one eruptive vent. Also, a volcano that has a dome an eruptive vent or vents (alt. 'complex', 'strato-volcano').

composite volcano A steep-pointed volcanic cone built of layers of ash, cinders, lava and other volcanic debris. Sometimes used interchangeably with the term 'strato-volcano'.

conduit The passage magma follows from within the Earth to reach the surface to erupt from a vent.

crater A bowl-shaped depression (usually smaller than 1.5 km / 1 mile) on a volcano's flanks or summit. Created either by collapse or an explosive eruption.

crust The outermost layer of Earth.

dacite Volcanic lava that is light in color and contains 62 to 69 per cent silica.

dome A hump that can form over a vent when very thick (viscous) doughy lava erupts and does not flow but builds up.

dormant Term used to refer to a sleeping, inactive volcano that may erupt again in the future.

eruption When solid, liquid, and gaseous matter is ejected into Earth's atmosphere or onto Earth's surface by volcanic activity. Eruptions can be gentle flows or devastating explosions.

extinct Used to describe a volcano that has not erupted in recorded history and may never erupt again.

fault A crack in the Earth's crust.

fissure Elongated cracks in the Earth's crust.

flood basalt Massive outpouring of basalt lava that covers portions of the Earth.

fumarole A vent or hole in a volcanic region emitting gas or steam.

hot spot A stationary plume of magma from deep inside Earth that burns through Earth's crust to feed active volcanoes.

kipuka A land island in a lava flow.

lahar A deadly mudflow formed of erupted volcanic materials mixed with water.

lava Molten rock that has reached the Earth's surface and erupted.

lava delta As hot molten lava flows into the ocean it hardens in a broad fan-shape of new land. Frequently molten lava still flows underneath.

magma Very hot, molten rock beneath the Earth's surface.

magma chamber A reservoir or storage area, that fills with magma under a volcano.

mantle The pliable zone between the Earth's crust and core.

nuée ardente See pyroclastic flow.

oceanic ridge A seam on the ocean floor where two bordering tectonic plates pull apart and lava erupts, creating new crust.

pahoehoe Hawaiian word, for smooth silica-poor extremely fluid basalt lava flows.

phreatic eruption An unexpected fierce eruption which occurs when water (glacial ice, lakes, rain, groundwater, oceans) and heated volcanic rocks come into contact with one another.

phreatomagmatic eruption Steam explosions caused by the interaction of rising magma and surface water.

pillow lava Bulbous, inflated, lobes of lava formed underwater by submarine eruptions and under ice.

plate Huge rigid rafts of crust material that comprises the Earth's outermost surface layer.

plate tectonics Scientific theory stating Earth's entire crust is fractured into about 12 sections or plates, which move away, toward, grind against, slip under and above each other causing volcanic activity.

plug Lava solidified in the conduit vent of a volcano.

pumice volcanic rock so saturated with gas bubbles that it resembles a sponge and floats on water.

pyroclasts Ash, pumice, lava, cinders and volcanic rocks ejected from a volcano.

pyroclastic flow A hurricane-force avalanche of hot gas, heated rocks, lava and ash that speeds down a volcano's flanks.

Pacific Ring of Fire An active region of earthquakes and volcanoes that surrounds the basin of the Pacific Ocean.

rhyolite Light colored volcanic rock or lava that contains 69 per cent silica or more.

sea floor spreading The creation of new crust on the sea floor when submarine plates pull away from each other.

seamount A submarine volcano.

shield volcano A wide, gently sloped volcano built of fluid basalt lava.

silica The glassy element in magma and lava that determines its viscosity. A chemical combination of silicon and oxygen. Silica-rich = thick. Silica-poor = thin.

strato-volcano See composite volcano.

subduction When one of Earth's tectonic plates is forced under another.

tephra A collective noun describing many types of volcanic materials that have been erupted up then deposited on the ground.

tsunami A giant ocean wave sometimes caused by volcanic collapse.

vent An opening in the Earth's crust where volcanic material is erupted from inside the Earth.

viscosity A liquid's resistance to flow (water has low viscosity while molasses has a higher viscosity.)

volcano a vent or opening in the Earth's surface and the subsequent cone formed around that vent, from accumulated eruptions of volcanic materials.

283

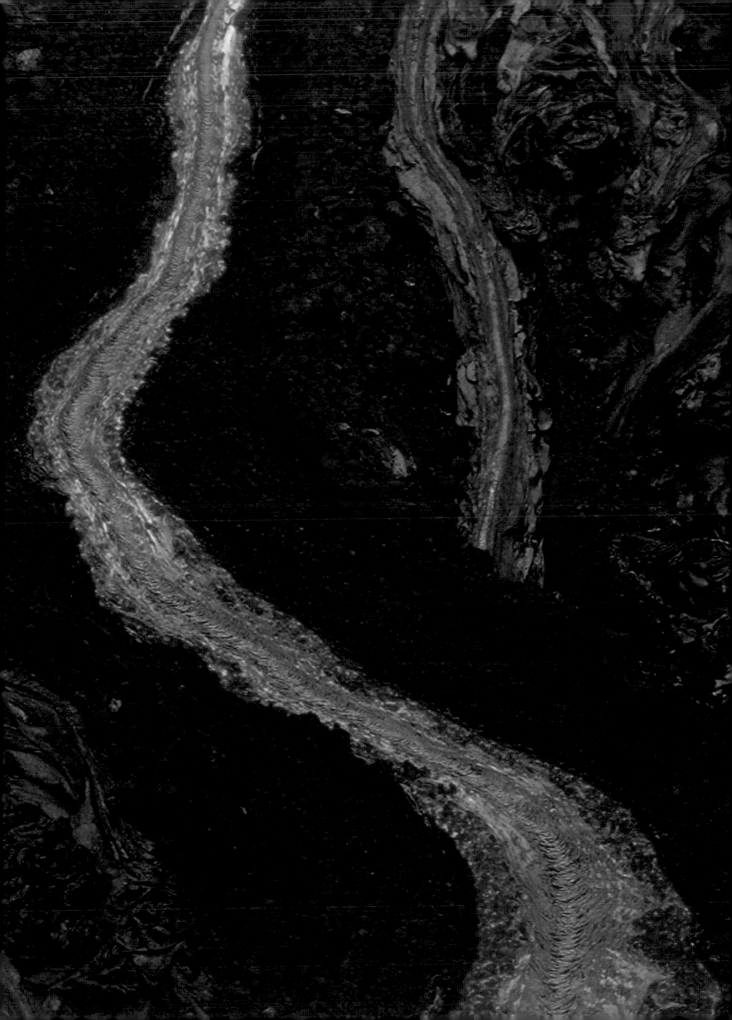

INDEX

285

287

ACKNOWLEDGMENTS

DEDICATION

This book is dedicated to my husband, Stephen James O'Meara, who has climbed with me to the mountain top and opened my eyes to the view.

AUTHOR'S ACKNOWLEDGMENTS

The climb to the summit is always easier with the help and support of other people. Volcano could not have been completed with out the sincere support of my husband Stephen James O'Meara, my loving family, dear Pele, Mandy, Milky way and Daisy, my patient and expert editor, Joanne Wilson, our awesome designer Austin, my wonderful agent Edward Necarsulmer IV, the friendly ears of Amos and Charlene Meyers and Brad Pitt.

PICTURE CREDITS

SPL=Science Photo Library

1 J.D. Griggs / Corbis; 2-3 Roger Ressmeyer / Corbis ; 4 Top: Bernhard Edmaier / SPL; Middle: Getty Images / Hideo Kurihara; Bottom: Whiddington / SPL; 5 Top: Galen Rowell / Corbis; Middle: Getty Images / G. Brad Lewis; Bottom: Corbis; 6-7 George Steinmetz / SPL; 9 Roger Ressmeyer / Corbis; 10-11 Stephen & Donna O'Meara / SPL; 12 James Anderson / Corbis; 13 Adam G. Sylvester / SPL; 14 Roger Ressmeyer / Corbis; 15 Jonathan Blair / Corbis; 16 Roger Ressmeyer / Corbis; 17 WEDA / epa / Corbis; 18 Left: Reuters / Corbis; 18-19 Hoa-Qui / Photo Researchers, Inc.; 19 Imelda Medina / epa / Corbis; 20 E. R. Degginger / SPL; 21 Jon Sparks / Corbis; 22 Hrafnsson Gisli Egill / Corbis Sygma; 23 Ric Ergenbright / Corbis; 24 National Geographical / Getty Images; 25 Alberto Garcia / Corbis; 26 Bernhard Edmaier / SPL; 27 Steve Kaufman / Corbis; 28-29 WEDA / epa / Corbis; 30 Earth Satellite Corporation / SPL; 31 Jaques Langevin / Corbis; 32 David Weintraub / SPL; 33 Loetscher Chlaus / Alamy; 34 NASA / SPL; 35 Sergio Dorantes / Corbis; 36 NASA / SPL; 37 Bettmann / Corbis; 38 Bernhard Edmaier / SPL; 39 NASA / SPL; 40-41 Bottom: Reuters / Corbis; Top: Getty Images / Richard A Cooke III; 42 Stephen & Donna O'Meara / SPL; 43 Cordaiy Photo Library Ltd. / Corbis; 44 NASA / SPL; 45 Jeremy Bishop / SPL; 46 Simon Fraser / SPL; 47 Top: Paul Souders / Corbis; Bottom: Bernhard Edmaier / SPL 48 Roger Ressmeyer / Corbis; 49 Peter & J Clement / SPL; 50 Stephen & Donna O'Meara / SPL; 51 Stephen & Donna O'Meara / SPL; 52 Roger Ressmeyer / Corbis; 53 Bettmann / Corbis; 54 Dr Juerg Alean / SPL; 55 Dr Juerg Alean / SPL; 56 Roger Ressmeyer / Corbis; 57 Pablo Corral Vega / Corbis; 58-59 Michele Falzone / JAI / Corbis; 60 Masao Hayashi / Duno / SPL; 61 Bernhard Edmaier / SPL; 62 Robert Gill; Papilio / Corbis; 63 David Parker / SPL; 64 Corbis; 65 Top: Corbis; Bottom:COLLART HERVE / Corbis SYGMA; 66-67 Mark A. Johnson / Corbis; Inset: Stephen & Donna O'Meara / SPL; 68 Bernardo De Niz / Reuters / Corbis; 69 Michael S. Yamashita / Corbis; 70 Carl & Ann Purcell / Corbis; 71 Dennis Sabangan / epa / Corbis; 72 Ric Ergenbright / Corbis; 73 Krafft / Hoa-Qui / SPL; 74 Douglas Peebles / Corbis; 75 CNES, 1991 Distribution Spot Image / SPL; 76 NASA / Corbis; 77 Profimedia International s.r.o. / Alamy; 78-79 Charles Mauzy / Corbis; 80 Stephen & Donna O'Meara SPL ; 81 J. Marshall,Tribaleye Images / Alamy; 82 Peter Johnson / Corbis; 83 Michele Falzone / JAI / Corbis; 84 Yann Arthus-Bertrand / Corbis; 85 Top: Brian A. Vikander / Corbis; Bottom: Earth Satellite Corporation / SPL; 86 Reuters / Corbis; 87 Corbis; 88 Jim Sugar / Corbis; 89 Top: M-Sat Ltd / SPL; Bottom: Staffan Widstrand / Corbis; 90 Yann Arthus-Bertrand / Corbis; 91 Jack Fields / Corbis; 92 Albrecht G. Schaefer / Corbis; 93 Yann Arthus-Bertrand / Corbis; 94 Bernhard Edmaier / SPL; 95 Corbis; 96 Getty Images / Hideo Kurihara; 97 Zephyr / SPL; 98 Nik Wheeler / Corbis; 99 Yann Arthus-Bertrand / Corbis; 100 Robert Harding Picture Library Ltd / Alamy; 101 Jose Azel (Contributor) Aurora / Getty Images; 102 Galen Rowell / Corbis; 103 Bernhard Edmaier / SPL; 104 Ria Novosti / SPL; 105 Jane Sweeney / Robert Harding World Imagery / Corbis; 106 Gary Braasch / Corbis; 107 Doug Beghtel / The Oregonian / Corbis; 108-109 Stephen & Donna O'Meara / SPL; 110 Reuters / Corbis; 111 José Fuste Raga / zefa / Corbis; 112 Stephen & Donna O'Meara SPL; 113 J.D. Griggs / Corbis; 114 Stephen & Donna O'Meara SPL; 115 Renate Jope SPL; 116 Richard A. Cooke / Corbis; 117 Gianni Tortoli / SPL; 118 Simon Fraser / SPL; 119 Stephen & Donna O'Meara / SPL; 120 Digital Globe. Eurimage / SPL; 121 Leonard Von Matt / SPL; 122 Getty Images / National Geographic / Carsen Peter; 124 Top: G. Brad Lewis / SPL; Bottom: G. Brad Lewis / SPL 125 Stephen & Donna O'Meara / SPL; 126 Matthew Shipp / SPL; 127 epa / Corbis; 128 Corbis; 129 Bettmann / Corbis; 130 Paul Almasy / Corbis; 131 Reuters / Corbis; 132 Corbis; 133 Stephen & Donna O'Meara / SPL; 134 Stephen & Donna O'Meara / SPL; 135 Stephen & Donna O'Meara / SPL; 136 C. Whiddington / SPL; 137 Mark Newman / SPL; 138 Corbis; 139 Top: Krafft / Photo Researchers, Inc.; Bottom: Explorer / Photo Researchers, Inc.140 Jonathan Blair / Corbis; 141 Christiana Carvalho; Frank Lane Picture Agency / Corbis; 142 Roger Ressmeyer / Corbis; 143 US Geological Survey / SPL; 144 Thierry Orban / Corbis SYGMA; 145 Bernhard Edmaier / SPL; 146 Bernhard Edmaier / SPL; 147 Jerome Minet / Kipa / Corbis; 148 R.L. Christiancen / SPL; 149 SPL; 150 Bernhard Edmaier / SPL; 151 Jane Sweeney / Robert Harding World Imagery / Corbis; 152 Dave G. Houser / Corbis; 153 Paul A. Souders / Corbis; 154 Roger Ressmeyer / Corbis; 155 Galen Rowell / Corbis; 156 Richard Glover / Corbis; 157 Bernhard Edmaier / SPL; 158 Jeremy Bishop / SPL; 159 Stephen & Donna O'Meara / SPL; 160 Stephen & Donna O'Meara / SPL; 161 Bernhard Edmaier / SPL; 162 G. Brad Lewis / SPL; 163 G. Brad Lewis / SPL; 164-165 Alan Sirulnikoff / SPL; 166 Stephen & Donna O'Meara / SPL; 167 E.R. Degginger / SPL; 168 SAT LTD / SPL; 169 Larry Dale Gordon / zefa / Corbis; 170 Bernhard Edmaier / SPL; 171 Dr Juerg Alean / SPL; 172 Pete Turner / Getty Images; 173 Charles Angelo / SPL; 174 Roger Ressmeyer / Corbis; 175 Tony Craddock / SPL; 176 Galen Rowell / Corbis, 177 Bo Zaunders / Corbis; 178 USGS / SPL; 179 Bernhard Edmaier / SPL; 180 Zephyr / SPL; 181 Zephyr / SPL ; 182 Dorothy Burrows; Eye Ubiquitous / Corbis; 183 Adam G. Sylvester / SPL; 184 Frank Lukasseck / Corbis; 185 NASA / SPL; 186 CNES, 2001 Distribution Spot Image / SPL; 187 David Muench / Corbis; 188 Dr Juerg Alean / SPL; 189 NASA / SPL; 190 Roger Ressmeyer / SPL; 191 Roger Ressmeyer/Corbis; 192 Phil Schermeister / Corbis; 193 Stephen & Donna O'Meara / SPL; 194-195 Douglas Peebles / Corbis; 196 NASA / SPL; 197 USGS / SPL; 198 Bernhard Edmaier / SPL; 199 Alexis Rosenfeld / SPL; 200 Jonathan Blair / Corbis; 201 Diane Cook & Len Jenshel / Corbis; 202 Dennis Flaherty / SPL; 203 David Parker / SPL; 204 Wolfgang Kaehler / Corbis; 205 Kenneth Murray / SPL; 206 CNES, Distribution Spot Image / SPL; 207 Shusuke Sezai / epa / Corbis; 208 A Sternberg / SPL; 209 British Antarctic Survey / SPL; 210-211 Reuters / Corbis; 212 Corbis; 213 Stephen & Donna O'Meara / SPL; 214-215 Roger Ressmeyer / Corbis; 216 Jack Fields / Corbis; 217 NASA / SPL; 218 Worldsat International / SPL; 219 Dietrich Rose / zefa / Corbis; 220 Kevin Schafer / Corbis; 221 Top: Institute Of Oceanographic Sciences / SPL; Bottom: USGS / SPL; 222 Corbis; 223 GEOEYE / SPL; 224 Getty Images / G. Brad Lewis; 225 Stephen & Donna O'Meara / SPL; 226-227 Getty Images / Paul Chesley / National Geographic; 228 Japan Coast Guard / Handout / Reuters / Corbis; 229 Massimo Sestini / Getty Images; 230 Simon Fraser / SPL; 231 Simon Fraser / SPL; 232 Japan Coast Guard / Handout / Reuters / Corbis; 233 Top: Fredrik Fransson / SPL; Bottom: Jesse Allen / NASA Earth Observatory / GSFC / MITI/ ERSDAC /JAROS / SPL; 234 Art Wolfe / SPL; 235 Ralph White / Corbis; 236 Jonathan Blair / Corbis; 237 SPL B. Murton / Southampton Oceanography Centre / SPL; 238 Ralph White / Corbis; 239 Pasquale Sorrentino / SPL; 240 B. Murton / Southampton Oceanography Centre / SPL; 241 Fred MsConnaghey / SPL; 242 Bernhard Edmaier / SPL; 243 NASA / SPL; 244 Layne Kennedy / Corbis ; 245 Ralph White / Corbis; 246 Top: Institute Of Oceanographic Sciences / NERC / SPL; Bottom: Dr. Ken MacDonald / SPL; 247 Bernhard Edmaier / SPL; 248 Danny Lehman / Corbis; 249 Pierre Vauthey / Corbis SYGMA; 250 NHPA / A.N.T. Photography; 251 G. Brad Lewis / SPL; 252-253 NASA / SPL; 254 Mariner 10 / Corbis; 255 NASA / JPL; 256 Corbis ; 257 NASA / Roger Ressmeyer / Corbis; 258 NASA / Roger Ressmeyer / Corbis; 259 NASA / Roger Ressmeyer / Corbis; 260 NASA / Roger Ressmeyer / Corbis; 261 Corbis; 262 NASA / Roger Ressmeyer / Corbis; 263 Roger Ressmeyer / Corbis; 264 Corbis; 265 NASA / SPL; 266-267 NASA / epa / Corbis; 267 NASA / SPL; 268 Corbis; 269 epa / Corbis; 270 NASA / JPL / Handout / Reuters / Corbis; 271 NASA / JPL / Cornell / ZUMA / Corbis; 272 NASA/ Roger Ressmeyer / Corbis; 273 USGS/ SPL; 274 NASA / Roger Ressmeyer / Corbis; 275 USGS / SPL; 276 NASA / Roger Ressmeyer / Corbis; 277 NASA / SPL; 278 Corbis; 279 NASA / JPL / Corbis; 280 NASA / CNP / Corbis; 281 Top: NASA / JPL / Corbis; Bottom: Roger Ressmeyer / Corbis; 282 Roger Ressmeyer / Corbis 284 Stephen & Donna O'Meara / SPL;

A FIREFLY BOOK

Published by Firefly Books Ltd. 2008

Text copyright © 2008 Donna O'Meara
Design and layout © 2008 Cassell Illustrated

First printing

Publisher Cataloging-in-Publication Data (U.S.)

O'Meara, Donna,1954–
    Volcano : spectacular images of a world on fire / Donna O'Meara.
[288] p. : col. photos. ; cm.
Includes index.
Summary: A journey to the myriad of volcanoes around the world and within the solar system.
ISBN-13: 978-1-55407-353-5 (pbk.)
ISBN-10:1-55407-353-7 (pbk.)
1. Volcanoes—Popular works. I. Title.
551.21 dc22   QE522.O443 2008

Library and Archives Canada Cataloguing in Publication

Donovan-O'Meara, Donna, 1954–
    Volcano : spectacular images of a world on fire / Donna O'Meara.
Includes index.
ISBN-13: 978-1-55407-353-5
ISBN-10:1-55407-353-7
    1. Volcanoes. I. Title.
QE522.D65 2008   551.21   C2007-904063-2

Published in the United States by
Firefly Books (U.S.) Inc.
P.O. Box 1338, Ellicott Station
Buffalo, New York 14205

Published in Canada by
Firefly Books Ltd.
66 Leek Crescent
Richmond Hill, Ontario L4B 1H1

Cover images
Front, Corbis/Kevin Schafer
Back, Corbis/Douglas Peebles

Printed in China